T0259434

# AN INTRODUCTION TO RHEOLOGY

RHEOLOGY SERIES

Advisory Editor: K. Walters FRS, Professor of Applied Mathematics,
University of Wales, Aberystwyth, U.K.

---

The photograph used on the cover shows an extreme example of the Weissenberg effect. It illustrates one of the most important non-linear effects in rheology, namely, the existence of normal stresses. The effect is produced, quite simply, by rotating a rod in a dish of viscoelastic liquid. The liquid in the photograph was prepared by dissolving a high molecular-weight polyisobutylene (Oppanol B200) in a low molecular-weight solvent of the same chemical nature (Hyvis 07, polybutene). As the rod rotates, the liquid climbs up it, whereas a Newtonian liquid would move towards the rim of the dish under the influence of inertia forces.

This particular experiment was set up and photographed at the Thornton Research Centre of Shell Research Ltd., and is published by their kind permission.

# AN INTRODUCTION TO RHEOLOGY

## H.A. Barnes

*Senior Scientist and Subject Specialist in Rheology and Fluid Mechanics,
Unilever Research, Port Sunlight Laboratory, Wirral, U.K.*

## J.F. Hutton

*Formerly: Principal Scientist and Leader of Tribology Group, Shell Research
Ltd., Ellesmere Port, U.K.*

*and*
## K. Walters

*Professor and Head of Department of Mathematics, University College of
Wales, Aberystwyth, U.K.*

Elsevier
Amsterdam – Lausanne – New York – Oxford – Shannon – Singapore – Tokyo

ELSEVIER B.V.
Radarweg 29
P.O. Box 211, 1000 AE
Amsterdam, The Netherlands

ELSEVIER Inc.
525 B Street
Suite 1900, San Diego
CA 92101-4495, USA

**ELSEVIER Ltd**
**The Boulevard**
**Langford Lane, Kidlington,**
**Oxford OX5 1GB, UK**

ELSEVIER Ltd
84 Theobalds Road
London WC1X 8RR
UK

First edition 1989
Second impression 1993
Third impresion 1996
Fourth impression 1997
Fifth impression 1998
Transferred to Digital Printing 2005

Library of Congress Cataloging in Publication Data

Barnes, H. A.
An introduction to rheology / by H.A. Barnes, J.F.Hutton, K. Walters.
    p.    cm.
 Bibliography: p.
Includes index.
ISBN 0-444087140-3 (U.S)
    1. Rheology.   I. Hutton, J. F. (John Fletcher), 1924-
II. Walters, Kenneth. III. Title.
QC189.5.B37  1989
532' . 51—dc19

British Library Cataloguing in Publication Data
A catalogue record is available from the British Library.


ISBN:      0-444-87140-3

Printed and bound by Antony Rowe Ltd, Eastbourne

# PREFACE

Rheology, defined as the *science of deformation and flow*, is now recognised as an important field of scientific study. A knowledge of the subject is essential for scientists employed in many industries, including those involving plastics, paints, printing inks, detergents, oils, etc. Rheology is also a respectable scientific discipline in its own right and may be studied by academics for their own esoteric reasons, with no major industrial motivation or input whatsoever.

The growing awareness of the importance of rheology has spawned a plethora of books on the subject, many of them of the highest class. It is therefore necessary at the outset to justify the need for yet another book.

Rheology is by common consent a difficult subject, and some of the necessary theoretical components are often viewed as being of prohibitive complexity by scientists without a strong mathematical background. There are also the difficulties inherent in any multidisciplinary science, like rheology, for those with a specific training e.g. in chemistry. Therefore, newcomers to the field are sometimes discouraged and for them the existing texts on the subject, some of which are outstanding, are of limited assistance on account of their depth of detail and highly mathematical nature.

For these reasons, it is our considered judgment after many years of experience in industry and academia, that there still exists a need for a modern *introductory* text on the subject; one which will provide an overview and at the same time ease readers into the necessary complexities of the field, pointing them at the same time to the more detailed texts on specific aspects of the subject.

In keeping with our overall objective, we have purposely (and with some difficulty) minimised the mathematical content of the earlier chapters and relegated the highly mathematical chapter on *Theoretical Rheology* to the end of the book. A glossary and bibliography are included.

A major component of the anticipated readership will therefore be made up of newcomers to the field, with at least a first degree or the equivalent in some branch of science or engineering (mathematics, physics, chemistry, chemical or mechanical engineering, materials science). For such, the present book can be viewed as an important (first) stepping stone on the journey towards a detailed appreciation of the subject with Chapters 1–5 covering foundational aspects of the subject and Chapters 6–8 more specialized topics. We certainly do not see ourselves in competition with existing books on rheology, and if this is not the impression gained on reading the present book we have failed in our purpose. We shall judge the success or otherwise of our venture by the response of newcomers to the field, especially those without a strong mathematical background. We shall not be unduly disturbed if long-standing rheologists find the book superficial, although we shall be deeply concerned if it is concluded that the book is unsound.

We express our sincere thanks to all our colleagues and friends who read earlier drafts of various parts of the text and made useful suggestions for improvement.

Mr Robin Evans is to be thanked for his assistance in preparing the figures and Mrs Pat Evans for her tireless assistance in typing the final manuscript.

H.A. Barnes
J.F. Hutton
K. Walters

# CONTENTS

CHAPTER 1

# INTRODUCTION

## 1.1 What is rheology?

The term 'Rheology' * was invented by Professor Bingham of Lafayette College, Indiana, on the advice of a colleague, the Professor of Classics. It means *the study of the deformation and flow of matter*. This definition was accepted when the American Society of Rheology was founded in 1929. That first meeting heard papers on the properties and behaviour of such widely differing materials as asphalt, lubricants, paints, plastics and rubber, which gives some idea of the scope of the subject and also the numerous scientific disciplines which are likely to be involved. Nowadays, the scope is even wider. Significant advances have been made in biorheology, in polymer rheology and in suspension rheology. There has also been a significant appreciation of the importance of rheology in the chemical processing industries. Opportunities no doubt exist for more extensive applications of rheology in the biotechnological industries. There are now national Societies of Rheology in many countries. The British Society of Rheology, for example, has over 600 members made up of scientists from widely differing backgrounds, including mathematics, physics, engineering and physical chemistry. In many ways, rheology has come of age.

## 1.2 Historical perspective

In 1678, Robert Hooke developed his *"True Theory of Elasticity"*. He proposed that "the power of any spring is in the same proportion with the tension thereof", i.e. *if you double the tension you double the extension*. This forms the basic premise behind the theory of classical (infinitesimal-strain) elasticity.

At the other end of the spectrum, Isaac Newton gave attention to liquids and in the *"Principia"* published in 1687 there appears the following hypothesis associated with the steady simple shearing flow shown in Fig. 1.1: "The resistance which arises from the lack of slipperiness of the parts of the liquid, other things being equal, is proportional to the velocity with which the parts of the liquid are separated from one another".

---

* Definitions of terms in single quotation marks are included in the Glossary.

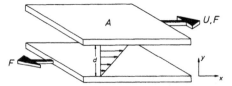

Fig. 1.1 Showing two parallel planes, each of area $A$, at $y = 0$ and $y = d$, the intervening space being filled with sheared liquid. The upper plane moves with relative velocity $U$ and the lengths of the arrows between the planes are proportional to the local velocity $v_x$ in the liquid.

This lack of slipperiness is what we now call 'viscosity'. It is synonymous with "internal friction" and is a measure of "resistance to flow". The force per unit area required to produce the motion is $F/A$ and is denoted by $\sigma$ and is proportional to the 'velocity gradient' (or 'shear rate') $U/d$, i.e. *if you double the force you double the velocity gradient*. The constant of proportionality $\eta$ is called the coefficient of viscosity, i.e.

$$\sigma = \eta U/d. \tag{1.1}$$

(It is usual to write $\dot{\gamma}$ for the shear rate $U/d$; see the Glossary.)

Glycerine and water are common liquids that obey Newton's postulate. For glycerine, the viscosity in SI units is of the order of 1 Pa.s, whereas the viscosity of water is about 1 mPa.s, i.e. one thousand times less viscous.

Now although Newton introduced his ideas in 1687, it was not until the nineteenth century that Navier and Stokes independently developed a consistent three-dimensional theory for what is now called a Newtonian viscous liquid. The governing equations for such a fluid are called the Navier–Stokes equations.

For the simple shear illustrated in Fig. 1.1, a 'shear stress' $\sigma$ results in 'flow'. In the case of a Newtonian liquid, the flow persists as long as the stress is applied. In contrast, for a Hookean solid, a shear stress $\sigma$ applied to the surface $y = d$ results in an instantaneous deformation as shown in Fig. 1.2. Once the deformed state is reached there is no further movement, but the deformed state persists as long as the stress is applied.

The angle $\gamma$ is called the 'strain' and the relevant 'constitutive equation' is

$$\sigma = G\gamma, \tag{1.2}$$

where $G$ is referred to as the 'rigidity modulus'.

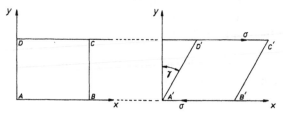

Fig. 1.2 The result of the application of a shear stress $\sigma$ to a block of Hookean solid (shown in section). On the application of the stress the material section ABCD is deformed and becomes A'B'C'D'.

Three hundred years ago everything may have appeared deceptively simple to Hooke and Newton, and indeed for two centuries everyone was satisfied with Hooke's Law for solids and Newton's Law for liquids. In the case of liquids, Newton's law was known to work well for some common liquids and people probably assumed that it was a universal law like his more famous laws about gravitation and motion. It was in the nineteenth century that scientists began to have doubts (see the review article by Markovitz (1968) for fuller details). In 1835, Wilhelm Weber carried out experiments on silk threads and found out that they were not perfectly elastic. "A longitudinal load", he wrote, "produced an immediate extension. This was followed by a further lengthening with time. On removal of the load an immediate contraction took place, followed by a gradual further decrease in length until the original length was reached". Here we have a solid-like material, whose behaviour cannot be described by Hooke's law alone. There are elements of flow in the described deformation pattern, which are clearly associated more with a liquid-like response. We shall later introduce the term 'viscoelasticity' to describe such behaviour.

So far as fluid-like materials are concerned, an influential contribution came in 1867 from a paper entitled "*On the dynamical theory of gases*" which appeared in the "*Encyclopaedia Britannica*". The author was James Clerk Maxwell. The paper proposed a mathematical model for a fluid possessing some elastic properties (see §3.3).

The definition of rheology already given would allow a study of the behaviour of all matter, including the classical extremes of Hookean elastic solids and Newtonian viscous liquids. However, these classical extremes are invariably viewed as being outside the scope of rheology. So, for example, *Newtonian fluid mechanics* based on the Navier–Stokes equations is not regarded as a branch of rheology and neither is classical *elasticity theory*. The over-riding concern is therefore with materials between these classical extremes, like Weber's silk threads and Maxwell's elastic fluids.

Returning to the historical perspective, we remark that the early decades of the twentieth century saw only the occasional study of rheological interest and, in general terms, one has to wait until the second World War to see rheology emerging as a force to be reckoned with. Materials used in flamethrowers were found to be viscoelastic and this fact generated its fair share of original research during the War. Since that time, interest in the subject has mushroomed, with the emergence of the synthetic-fibre and plastics-processing industries, to say nothing of the appearance of liquid detergents, multigrade oils, non-drip paints and contact adhesives. There have been important developments in the pharmaceutical and food industries and modern medical research involves an important component of biorheology. The manufacture of materials by the biotechnological route requires a good understanding of the rheology involved. All these developments and materials help to illustrate the substantial relevance of rheology to life in the second half of the twentieth century.

### 1.3 The importance of non-linearity

So far we have considered elastic behaviour and viscous behaviour in terms of the laws of Hooke and Newton. These are linear laws, which assume direct proportionality between stress and strain, or strain rate, whatever the stress. Further, by implication, the viscoelastic behaviour so far considered is also linear. Within this linear framework, a wide range of rheological behaviour can be accommodated. However, this framework is very restrictive. The range of stress over which materials behave linearly is invariably limited, and the limit can be quite low. In other words, material properties such as rigidity modulus and viscosity can change with the applied stress, and the stress need not be high. The change can occur either instantaneously or over a long period of time, and it can appear as either an increase or a decrease of the material parameter.

A common example of non-linearity is known as 'shear-thinning' (cf. §2.3.2). This is a reduction of the viscosity with increasing rate of shear in steady flow. The toothpaste which sits apparently unmoving on the bristles of the toothbrush is easily squeezed from the toothpaste tube—a familiar example of shear-thinning. The viscosity changes occur almost instantaneously in toothpaste. For an example of shear-thinning which does not occur instantaneously we look to non-drip paint. To the observer equipped with no more than a paintbrush the slow recovery of viscosity is particularly noticeable. The special term for time-dependent shear-thinning followed by recovery is 'thixotropy', and non-drip paint can be described as thixotropic. Shear-thinning is just one manifestation of non-linear behaviour, many others could be cited, and we shall see during the course of this book that it is difficult to make much headway in the understanding of rheology without an appreciation of the general importance of non-linearity.

### 1.4 Solids and liquids

It should now be clear that the concepts of elasticity and viscosity need to be qualified since real materials can be made to display either property or a combination of both simultaneously. Which property dominates, and what the values of the parameters are, depend on the stress and the duration of application of the stress.

The reader will now ask what effect these ideas will have on the even more primitive concepts of solids and liquids. The answer is that in a detailed discussion of real materials these too will need to be qualified. When we look around at home, in the laboratory, or on the factory floor, we recognise solids or liquids by their response to low stresses, usually determined by gravitational forces, and over a human, everyday time-scale, usually no more than a few minutes or less than a few seconds. However, if we apply a very wide range of stress over a very wide spectrum of time, or frequency, using rheological apparatus, we are able to observe liquid-like properties in solids and solid-like properties in liquids. It follows therefore that difficulties can, and do, arise when an attempt is made to label a given material as a solid or a liquid. In fact, we can go further and point to inadequacies even when qualifying terms are used. For example, the term plastic–rigid solid used in structural engineering to denote a material which is rigid (inelastic) below a 'yield stress' and yielding indefinitely above this stress, is a good approximation for a structural component of a steel bridge but it is nevertheless still limited as a description for steel. It is much more fruitful to classify rheological *behaviour*. Then it will be possible to include a given material in more than one of these classifications depending on the experimental conditions.

A great advantage of this procedure is that it allows for the mathematical description of rheology as the mathematics of a set of behaviours rather than of a set of materials. The mathematics then leads to the proper definition of rheological parameters and therefore to their proper measurement (see also §3.1).

To illustrate these ideas, let us take as an example, the silicone material that is nicknamed "Bouncing Putty". It is very viscous but it will eventually find its own level when placed in a container—given sufficient time. However, as its name suggests, a ball of it will also bounce when dropped on the floor. It is not difficulty to conclude that in a slow flow process, occurring over a long time scale, the putty behaves like a liquid—it finds its own level slowly. Also when it is extended slowly it shows ductile fracture—a liquid characteristic. However, when the putty is extended quickly, i.e. on a shorter time scale, it shows brittle fracture—a solid characteristic. Under the severe and sudden deformation experienced as the putty strikes the ground, it bounces—another solid characteristic. Thus, a given material can behave like a solid or a liquid depending on the time scale of the deformation process.

The scaling of time in rheology is achieved by means of the 'Deborah number', which was defined by Professor Marcus Reiner, and may be introduced as follows.

Anyone with a knowledge of the QWERTY keyboard will know that the letter "R" and the letter "T" are next to each other. One consequence of this is that any book on *rheology* has at least one incorrect reference to *theology*. (Hopefully, the present book is an exception!). However, this is not to say that there is no connection between the two. In the fifth chapter of the book of Judges in the Old Testament, Deborah is reported to have declared, "The mountains flowed before the Lord...". On the basis of this reference, Professor Reiner, one of the founders of the modern science of rheology, called his dimensionless group the Deborah number $D_e$. The idea is that *everything flows if you wait long enough*, even the mountains!

$$D_e = \tau/T,\tag{1.3}$$

where $T$ is a characteristic time of the deformation process being observed and $\tau$ is a characteristic time of the material. The time $\tau$ is infinite for a Hookean elastic solid and zero for a Newtonian viscous liquid. In fact, for water in the liquid state $\tau$ is typically $10^{-12}$ s whilst for lubricating oils as they pass through the high pressures encountered between contacting pairs of gear teeth $\tau$ can be of the order of $10^{-6}$ s and for polymer melts at the temperatures used in plastics processing $\tau$ may be as high as a few seconds. There are therefore situations in which these liquids depart from purely viscous behaviour and also show elastic properties.

High Deborah numbers correspond to solid-like behaviour and low Deborah numbers to liquid-like behaviour. A material can therefore appear solid-like either because it has a very long characteristic time or because the deformation process we are using to study it is very fast. Thus, even mobile liquids with low characteristic times can behave like elastic solids in a very fast deformation process. This sometimes happens when lubricating oils pass through gears.

Notwithstanding our stated decision to concentrate on material *behaviour*, it may still be helpful to attempt definitions of precisely what we mean by *solid* and *liquid*, since we do have recourse to refer to such expressions in this book. Accordingly, we define a solid as *a material that will not continuously change its shape when subjected to a given stress*, i.e. for a given stress there will be a fixed final deformation, which may or may not be reached instantaneously on application of the stress. We define a liquid as *a material that will continuously change its shape (i.e. will flow) when subjected to a given stress, irrespective of how small that stress may be*.

The term 'viscoelasticity' is used to describe behaviour which falls between the classical extremes of Hookean elastic response and Newtonian viscous behaviour. In terms of ideal material response, a solid material with viscoelasticity is invariably called a 'viscoelastic solid' in the literature. In the case of liquids, there is more ambiguity so far as terminology is concerned. The terms 'viscoelastic liquid', 'elastico-viscous liquid', 'elastic liquid' are all used to describe a liquid showing viscoelastic properties. In recent years, the term 'memory fluid' has also been used in this connection. In this book, we shall frequently use the simple term elastic liquid.

Liquids whose behaviour cannot be described on the basis of the Navier–Stokes equations are called 'non-Newtonian liquids'. Such liquids may or may not possess viscoelastic properties. This means that all viscoelastic liquids are non-Newtonian, but the converse is not true: not all non-Newtonian liquids are viscoelastic.

## 1.5 Rheology is a difficult subject

By common consent, rheology is a difficult subject. This is certainly the usual perception of the newcomer to the field. Various reasons may be put forward to explain this view. For example, the subject is interdisciplinary and most scientists and engineers have to move away from a possibly restricted expertise and develop a broader scientific approach. The theoretician with a background in continuum mechanics needs to develop an appreciation of certain aspects of physical chemistry, statistical mechanics and other disciplines related to microrheological studies to fully appreciate the breadth of present-day rheological knowledge. Even more daunting, perhaps, is the need for non-mathematicians to come to terms with at least some aspects of non-trivial mathematics. A cursory glance at most text books on rheology would soon convince the uninitiated of this. Admittedly, the apparent need of a working knowledge of such subjects as *functional analysis* and *general tensor analysis* is probably overstated, but there is no doubting the requirement of some working knowledge of modern mathematics. This book is an *introduction* to rheology and our stated aim is to explain any mathematical complication to the nonspecialist. We have tried to keep to this aim throughout most of the book (until Chapter 8, which is written for the more mathematically minded reader).

At this point, we need to justify the introduction of the indicial notation, which is an essential mathematical tool in the development of the subject. The concept of *pressure* as a (normal) force per unit area is widely accepted and understood; it is taken for granted, for example, by TV weather forecasters who are happy to display isobars on their weather maps. Pressure is viewed in these contexts as a scalar quantity, but the move to a more sophisticated (tensor) framework is necessary when viscosity and other rheological concepts are introduced.

We consider a small plane surface of area $\Delta s$ drawn in a deforming medium (Fig. 1.3).

Let $n_x$, $n_y$ and $n_z$ represent the components of the unit normal vector to the surface in the $x$, $y$, $z$ directions, respectively. These define the orientation of $\Delta s$ in space. The normal points in the direction of the +ve side of the surface. We say that the material on the +ve side of the surface exerts a force with components $F_x^{(n)} \Delta s$, $F_y^{(n)} \Delta s$, $F_z^{(n)} \Delta s$ on the material on the −ve side, it being implicitly assumed that the area $\Delta s$ is small enough for the 'stress' components $F_x^{(n)}$, $F_y^{(n)}$, $F_z^{(n)}$ to be regarded as constant over the small surface $\Delta s$. A more convenient notation is to replace these components by the stress components $\sigma_{nx}$, $\sigma_{ny}$, $\sigma_{nz}$, the first index referring to the *orientation* of the plane surface and the second to the *direction* of the stress. Our sign convention, which is universally accepted, except by Bird et al. (1987(a) and (b)), is that a positive $\sigma_{zz}$ (and similarly $\sigma_{yy}$ and $\sigma_{xx}$) is a tension. Components $\sigma_{xx}$, $\sigma_{yy}$ and $\sigma_{zz}$ are termed 'normal stresses' and $\sigma_{zx}$, $\sigma_{zy}$ etc. are called 'shear stresses'. It may be formally shown that $\sigma_{xy} = \sigma_{yx}$, $\sigma_{xz} = \sigma_{zx}$ and $\sigma_{yz} = \sigma_{zy}$ (see, for example Schowalter 1978, p. 44).

Figure 1.4 may be helpful to the newcomer to continuum mechanics to explain

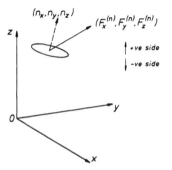

Fig. 1.3 The mutually perpendicular axes $0x$, $0y$, $0z$ are used to define the position and orientation of the small area $\Delta s$ and the force on it.

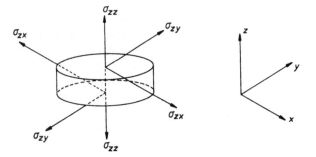

Fig. 1.4 The components of stress on the plane surfaces of a volume element of a deforming medium.

the relevance of the indicial notation. The figure contains a schematic representation of the stress components on the plane surfaces of a small volume which forms part of a general continuum. The stresses shown are those acting *on* the small volume *due* to the surrounding material.

The need for an indicial notation is immediately illustrated by a more detailed consideration of the steady simple-shear flow associated with Newton's postulate (Fig. 1.1), which we can conveniently express in the mathematical form:

$$v_x = \dot{\gamma}y, \quad v_y = v_z = 0, \tag{1.4}$$

where $v_x$, $v_y$, $v_z$ are the velocity components in the $x$, $y$ and $z$ directions, respectively, and $\dot{\gamma}$ is the (constant) shear rate. In the case of a Newtonian liquid, the stress distribution for such a flow can be written in the form

$$\sigma_{yx} = \eta\dot{\gamma}, \quad \sigma_{xz} = \sigma_{yz} = 0, \quad \sigma_{xx} - \sigma_{yy} = 0, \quad \sigma_{yy} - \sigma_{zz} = 0, \tag{1.5}$$

and here there would be little purpose in considering anything other than the shear stress $\sigma_{yx}$ which we wrote as $\sigma$ in eqn. (1.1). Note that it is usual to work in terms of normal stress *differences* rather than the normal stresses themselves, since the latter are arbitrary to the extent of an added isotropic pressure in the case of incompressible liquids and we would need to replace (1.5) by

$$\left.\begin{array}{l} \sigma_{yx} = \eta\dot{\gamma}, \quad \sigma_{xz} = \sigma_{yz} = 0, \\ \sigma_{xx} = -p, \quad \sigma_{yy} = -p, \quad \sigma_{zz} = -p, \end{array}\right\} \tag{1.6}$$

where $p$ is an arbitrary isotropic pressure. There is clearly merit in using (1.5) rather than (1.6) since the need to introduce $p$ is avoided (see also Dealy 1982, p. 8).

For elastic liquids, we shall see in later chapters that the stress distribution is more complicated, requiring us to modify (1.5) in the following manner:

$$\left.\begin{array}{l} \sigma_{yx} = \eta(\dot{\gamma})\dot{\gamma}, \quad \sigma_{xz} = \sigma_{yz} = 0, \\ \sigma_{xx} - \sigma_{yy} = N_1(\dot{\gamma}), \quad \sigma_{yy} - \sigma_{zz} = N_2(\dot{\gamma}), \end{array}\right\} \tag{1.7}$$

where it is now necessary to allow the viscosity to vary with shear rate, written mathematically as the function $\eta(\dot{\gamma})$, and to allow the normal stresses to be non-zero and also functions of $\dot{\gamma}$. Here the so called normal stress differences $N_1$ and $N_2$ are of significant importance and it is difficult to see how they could be conveniently introduced without an indicial notation *. Such a notation is therefore not an optional extra for mathematically-minded researchers but an absolute necessity. Having said that, we console non-mathematical readers with the promise that this represents the only major mathematical difficulty we shall meet until we tackle the notoriously difficult subject of constitutive equations in Chapter 8.

## 1.6 Components of rheological research

Rheology is studied by both university researchers and industrialists. The former may have esoteric as well as practical reasons for doing so, but the industrialist, for obvious reasons, is driven by a more pragmatic motivation. But, whatever the background or motivation, workers in rheology are forced to become conversant with certain well-defined sub-areas of interest which are detailed below. These are (i) rheometry; (ii) constitutive equations; (iii) measurement of flow behaviour in (non-rheometric) complex geometries; (iv) calculation of behaviour in complex flows.

### 1.6.1 Rheometry

In 'rheometry', materials are investigated in simple flows like the steady simple-shear flow already discussed. It is an important component of rheological research. Small-amplitude oscillatory-shear flow (§3.5) and extensional flow (Chapter 5) are also important.

The motivation for any rheometrical study is often the hope that observed behaviour in industrial situations can be correlated with some easily measured rheometrical function. Rheometry is therefore of potential importance in quality control and process control. It is also of potential importance in assessing the usefulness of any proposed constitutive model for the test material, whether this is based on molecular or continuum ideas. Indirectly, therefore, rheometry may be relevant in industrial process modelling. This will be especially so in future when the full potential of *computational fluid dynamics* using large computers is realized within a rheological context.

A number of detailed texts dealing specifically with rheometry are available. These range from the "How to" books of Walters (1975) and Whorlow (1980) to the "Why?" books of Walters (1980) and Dealy (1982). Also, most of the standard texts

---

* By common convention $N_1$ is called the first normal stress difference and $N_2$ the second normal stress difference. However, the terms "primary" and "secondary" are also used. In some texts $N_1$ is defined as $\sigma_{xx} - \sigma_{zz}$, whilst $N_2$ remains as $\sigma_{yy} - \sigma_{zz}$.

on rheology contain a significant element of rheometry, most notably Lodge (1964, 1974), Bird et al. (1987(a) and (b)), Schowalter (1978), Tanner (1985) and Janeschitz-Kriegl (1983). This last text also considers 'flow birefringence', which will not be discussed in detail in the present book (see also Doi and Edwards 1986, §4.7).

### 1.6.2 Constitutive equations

Constitutive equations (or rheological equations of state) are equations relating suitably defined stress and deformation variables. Equation (1.1) is a simple example of the relevant constitutive law for the Newtonian viscous liquid.

Constitutive equations may be derived from a microrheological standpoint, where the molecular structure is taken into account explicitly. For example, the solvent and polymer molecules in a polymer solution are seen as distinct entities. In recent years there have been many significant advances in microrheological studies.

An alternative approach is to take a continuum (macroscopic) point of view. Here, there is no direct appeal to the individual microscopic components, and, for example, a polymer solution is treated as a homogeneous continuum.

The basic discussion in Chapter 8 will be based on the principles of *continuum mechanics*. No attempt will be made to give an all-embracing discourse on this difficult subject, but it is at least hoped to point the interested and suitably equipped reader in the right direction. Certainly, an attempt will be made to assess the status of the more popular constitutive models that have appeared in the literature, whether these arise from microscopic or macroscopic considerations.

### 1.6.3 Complex flows of elastic liquids

The flows used in rheometry, like the viscometric flow shown in Fig. 1.1, are generally regarded as being *simple* in a rheological sense. By implication, all other flows are considered to be *complex*. Paradoxically, complex flows can sometimes occur in what appear to be simple geometrical arrangements, e.g. flow into an abrupt contraction (see §5.4.6). The complexity in the flow usually arises from the coexistence of shear and extensional components; sometimes with the added complication of inertia. Fortunately, in many cases, complex flows can be dealt with by using various numerical techniques and computers.

The experimental and theoretical study of the behaviour of elastic liquids in complex flows is generating a significant amount of research at the present time. In this book, these areas will not be discussed in detail: they are considered in depth in recent review articles by Boger (1987) and Walters (1985); and the important subject of the *numerical simulation* of non-Newtonian flow is covered by the text of Crochet, Davies and Walters (1984).

# CHAPTER 2

# VISCOSITY

## 2.1 Introduction

In this chapter we give detailed attention to the steady shear viscosity which was introduced in Chapter 1. Such a study is appropriate, since viscosity is traditionally regarded as a most important material property and any practical study requiring a knowledge of material response would automatically turn to the viscosity in the first instance.

The concept of viscosity was introduced in §1.2 through Newton's postulate, in which the shear stress $\sigma$ was related to the velocity gradient, or shear rate $\dot{\gamma}$, through the equation

$$\sigma = \eta\dot{\gamma}, \tag{2.1}$$

where $\eta$ is the shear viscosity. Table 2.1 is an approximate guide to the range of viscosities of familiar materials at room temperature and pressure. Most of the examples shown exhibit Newtonian behaviour under normal circumstances, by which we mean that $\eta$ is independent of shear rate for the shear rates of interest.

For Newtonian liquids, $\eta$ is sometimes called the coefficient of viscosity but it is now more commonly referred to simply as *the viscosity*. Such a terminology is helpful within the context of rheology, since, for most liquids, $\eta$ is not a *coefficient*,

TABLE 2.1
The viscosity of some familiar materials at room temperature

| Liquid | Approximate viscosity (Pa.s) |
|---|---|
| Glass | $10^{40}$ |
| Molten glass (500 ° C) | $10^{12}$ |
| Bitumen | $10^{8}$ |
| Molten polymers | $10^{3}$ |
| Golden syrup | $10^{2}$ |
| Liquid honey | $10^{1}$ |
| Glycerol | $10^{0}$ |
| Olive oil | $10^{-1}$ |
| Bicycle oil | $10^{-2}$ |
| Water | $10^{-3}$ |
| Air | $10^{-5}$ |

11

but a *function* of the shear rate $\dot{\gamma}$. We define the function $\eta(\dot{\gamma})$ as the 'shear viscosity' or simply viscosity, although in the literature it is often referred to as the 'apparent viscosity' or sometimes as the shear-dependent viscosity. An instrument designed to measure viscosity is called a 'viscometer'. A viscometer is a special type of 'rheometer' (defined as an instrument for measuring rheological properties) which is limited to the measurement of viscosity.

The current SI unit of viscosity is the Pascal-second which is abbreviated to Pa.s. Formerly, the widely used unit of viscosity in the cgs system was the Poise, which is smaller than the Pa.s by a factor of 10. Thus, for example, the viscosity of water at 20.2°C is 1 mPa.s (milli-Pascal second) and was 1 cP (centipoise).

In the following discussion we give a general indication of the relevance of viscosity to a number of practical situations; we discuss its measurement using various viscometers; we also study its variation with such experimental conditions as shear rate, time of shearing, temperature and pressure.

## 2.2 Practical ranges of variables which affect viscosity

The viscosity of real materials can be significantly affected by such variables as shear rate, temperature, pressure and time of shearing, and it is clearly important for us to highlight the way viscosity depends on such variables. To facilitate this, we first give a brief account of viscosity changes observed over practical ranges of interest of the main variables concerned, before considering in depth the shear rate, which from the rheological point of view, is the most important influence on viscosity.

### 2.2.1 Variation with shear rate

Table 2.2 shows the approximate magnitude of the shear rates encountered in a number of industrial and everyday situations in which viscosity is important and therefore needs to be measured. The approximate shear rate involved in any operation can be estimated by dividing the average velocity of the flowing liquid by a characteristic dimension of the geometry in which it is flowing (e.g. the radius of a tube or the thickness of a sheared layer). As we see from Table 2.2, such calculations for a number of important applications give an enormous range, covering 13 orders of magnitude form $10^{-6}$ to $10^7 \text{ s}^{-1}$. Viscometers can now be purchased to measure viscosity over this entire range, but at least three different instruments would be required for the purpose.

In view of Table 2.2, it is clear that the shear-rate dependence of viscosity is an important consideration and, from a practical standpoint, it is as well to have the particular application firmly in mind before investing in a commercial viscometer.

We shall return to the shear-rate dependence of viscosity in §2.3.

### 2.2.2 Variation with temperature

So far as temperature is concerned, for most industrial applications involving aqueous systems, interest is confined to 0 to 100°C. Lubricating oils and greases are

TABLE 2.2
Shear rates typical of some familiar materials and processes

| Situation | Typical range of shear rates (s$^{-1}$) | Application |
|---|---|---|
| Sedimentation of fine powders in a suspending liquid | $10^{-6}$–$10^{-4}$ | Medicines, paints |
| Levelling due to surface tension | $10^{-2}$–$10^{-1}$ | Paints, printing inks |
| Draining under gravity | $10^{-1}$–$10^{1}$ | Painting and coating. Toilet bleaches |
| Extruders | $10^{0}$–$10^{2}$ | Polymers |
| Chewing and swallowing | $10^{1}$–$10^{2}$ | Foods |
| Dip coating | $10^{1}$–$10^{2}$ | Paints, confectionary |
| Mixing and stirring | $10^{1}$–$10^{3}$ | Manufacturing liquids |
| Pipe flow | $10^{0}$–$10^{3}$ | Pumping. Blood flow |
| Spraying and brushing | $10^{3}$–$10^{4}$ | Spray-drying, painting, fuel atomization |
| Rubbing | $10^{4}$–$10^{5}$ | Application of creams and lotions to the skin |
| Milling pigments in fluid bases | $10^{3}$–$10^{5}$ | Paints, printing inks |
| High speed coating | $10^{5}$–$10^{6}$ | Paper |
| Lubrication | $10^{3}$–$10^{7}$ | Gasoline engines |

used from about $-50\,°C$ to $300\,°C$. Polymer melts are usually handled in the range $150\,°C$ to $300\,°C$, whilst molten glass is processed at a little above $500\,°C$.

Most of the available laboratory viscometers have facilities for testing in the range $-50\,°C$ to $150\,°C$ using an external temperature controller and a circulating fluid or an immersion bath. At higher temperatures, air baths are used.

The viscosity of Newtonian liquids decreases with increase in temperature, approximately according to the Arrhenius relationship:

$$\eta = A\mathrm{e}^{-B/T}, \tag{2.2}$$

where $T$ is the absolute temperature and $A$ and $B$ are constants of the liquid. In general, for Newtonian liquids, the greater the viscosity, the stronger is the temperature dependence. Figure 2.1 shows this trend for a number of lubricating oil fractions.

The strong temperature dependence of viscosity is such that, to produce accurate results, great care has to be taken with temperature control in viscometry. For instance, the temperature sensitivity for water is 3% per $°C$ at room temperature, so that $\pm1\%$ accuracy requires the sample temperature to be maintained to within $\pm0.3\,°C$. For liquids of higher viscosity, given their stronger viscosity dependence on temperature, even greater care has to be taken.

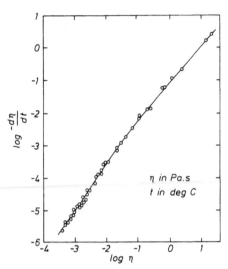

Fig. 2.1. Logarithm of viscosity/temperature derivative versus logarithm of viscosity for various lubricating oil fractions (Cameron 1966, p. 27).

It is important to note that it is not sufficient in viscometry to simply maintain control of the thermostat temperature; the act of shearing itself generates heat within the liquid and may thus change the temperature enough to decrease the viscosity, unless steps are taken to remove the heat generated. The rate of energy dissipation per unit volume of the sheared liquid is the product of the shear stress and shear rate or, equivalently, the product of the viscosity and the square of the shear rate.

Another important factor is clearly the rate of heat extraction, which in viscometry depends on two things. First, the kind of apparatus: in one class the test liquid flows through and out of the apparatus whilst, in the other, test liquid is permanently contained within the apparatus. In the first case, for instance in slits and capillaries, the liquid flow itself convects some of the heat away. On the other hand, in instruments like the concentric cylinder and cone-and-plate viscometers, the conduction of heat to the surfaces is the only significant heat-transfer process.

Secondly, heat extraction depends on the dimensions of the viscometers: for slits and capillaries the channel width is the controlling parameter, whilst for concentric cylinders and cone-and-plate devices, the gap width is important. It is desirable that these widths be made as small as possible.

### 2.2.3 Variation with pressure

The viscosity of liquids increases exponentially with isotropic pressure. Water below 30 °C is the only exception, in which case it is found that the viscosity first decreases before eventually increasing exponentially. The changes are quite small for pressures differing from atmospheric pressure by about one bar. Therefore, for most practical purposes, the pressure effect is ignored by viscometer users. There

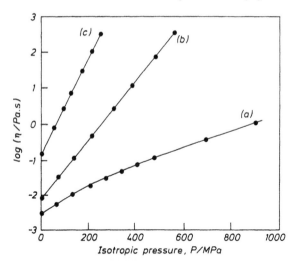

Fig. 2.2. Variation of viscosity with pressure: (a) Di–(2-ethylhexyl) sebacate; (b) Naphthenic mineral oil at 210 °F; (c) Naphthenic mineral oil at 100 °F. (Taken from Hutton 1980.)

are, however, situations where this would not be justified. For example, the oil industry requires measurements of the viscosity of lubricants and drilling fluids at elevated pressures. The pressures experienced by lubricants in gears can often exceed 1 GPa, whilst oil-well drilling muds have to operate at depths where the pressure is about 20 MPa. Some examples of the effect of pressure on lubricants is given in Fig. 2.2 where it can be seen that a viscosity rise of four orders of magnitude can occur for a pressure rise from atmospheric to 0.5 GPa.

## 2.3 The shear-dependent viscosity of non-Newtonian liquids

### 2.3.1 Definition of Newtonian behaviour

Since we shall concentrate on *non*-Newtonian viscosity behaviour in this section, it is important that we first emphasize what Newtonian behaviour is, in the context of the shear viscosity.

Newtonian behaviour in experiments conducted at constant temperature and pressure has the following characteristics:

*(i)* The only stress generated in simple shear flow is the shear stress $\sigma$, the two normal stress differences being zero.

*(ii)* The shear viscosity does not vary with shear rate.

*(iii)* The viscosity is constant with respect to the time of shearing and the stress in the liquid falls to zero immediately the shearing is stopped. In any subsequent

shearing, however long the period of resting between measurements, the viscosity is as previously measured.

*(iv)* The viscosities measured in different types of deformation are always in simple proportion to one another, so, for example, the viscosity measured in a uniaxial extensional flow is always three times the value measured in simple shear flow (cf. §5.3).

A liquid showing any deviation from the above behaviour is non-Newtonian.

### 2.3.2 The shear-thinning non-Newtonian liquid

As soon as viscometers became available to investigate the influence of shear rate on viscosity, workers found departure from Newtonian behaviour for many materials such as dispersions, emulsions and polymer solutions. In the vast majority of cases, the viscosity was found to decrease with increase in shear rate, giving rise to what is now generally called 'shear-thinning' behaviour although the terms temporary viscosity loss and 'pseudoplasticity' have also been employed.*

We shall see that there are cases (albeit few in number) where the viscosity increases with shear rate. Such behaviour is generally called 'shear-thickening' although the term 'dilatancy' has also been used.

For shear-thinning materials, the general shape of the curve representing the variation of viscosity with shear stress is shown in Fig. 2.3. The corresponding graphs of shear stress against shear rate and viscosity against shear rate are also given.

The curves indicate that in the limit of very low shear rates (or stresses) the viscosity is constant, whilst in the limit of high shear rates (or stresses) the viscosity is again constant, but at a lower level. These two extremes are sometimes known as the lower and upper Newtonian regions, respectively, the *lower* and *upper* referring to the *shear rate* and not the viscosity. The terms "first Newtonian region" and "second Newtonian region" have also been used to describe the two regions where the viscosity reaches constant values. The higher constant value is called the "zero-shear viscosity".

Note that the liquid of Fig. 2.3 does not show '*yield stress*' behaviour although if the experimental range had been $10^4$ s$^{-1}$ to $10^{-1}$ s$^{-1}$ (which is quite a wide range) an interpretation of the modified Fig. 2.3(b) might draw that conclusion. In Fig. 2.3(b) we have included so-called 'Bingham' plastic behaviour for comparison purposes. By definition, Bingham plastics will not flow until a critical yield stress $\sigma_y$ is exceeded. Also, by implication, the viscosity is infinite at zero shear rate and there is no question of a first Newtonian region in this case.

There is no doubt that the concept of yield stress can be helpful in some practical situations, but the question of whether or not a yield stress exists or whether all non-Newtonian materials will exhibit a finite zero-shear viscosity becomes of more

---

* The German word is "strukturviscosität" which is literally translated as *structural viscosity*, and is not a very good description of shear-thinning.

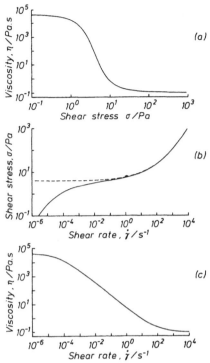

Fig. 2.3. Typical behaviour of a non-Newtonian liquid showing the interrelation between the different parameters. The same experimental data are used in each curve. (a) Viscosity versus shear stress. Notice how fast the viscosity changes with shear stress in the middle of the graph; (b) Shear stress versus shear rate. Notice that, in the middle of the graph, the stress changes very slowly with increasing shear rate. The dotted line represents ideal yield-stress (or Bingham plastic) behaviour; (c) Viscosity versus shear rate. Notice the wide range of shear rates needed to traverse the entire flow curve.

than esoteric interest as the range and sophistication of modern constant-stress viscometers make it possible to study the very low shear-rate region of the viscosity curve with some degree of precision (cf. Barnes and Walters 1985). We simply remark here that for dilute solutions and suspensions, there is no doubt that flow occurs at the smallest stresses: the liquid surface levels out under gravity and there is no yield stress. For more concentrated systems, particularly for such materials as gels, lubricating greases, ice cream, margarine and stiff pastes, there is understandable doubt as to whether or not a yield stress exists. It is easy to accept that a lump of one of these materials will never level out under its own weight. Nevertheless there is a growing body of experimental evidence to suggest that even concentrated systems flow in the limit of very low stresses. These materials appear not to flow merely because the zero shear viscosity is so high. If the viscosity is $10^{10}$ Pa.s it would take years for even the slightest flow to be detected visually!

The main factor which now enables us to explore with confidence the very low shear-rate part of the viscosity curve is the availability, on a commercial basis, of constant stress viscometers of the Deer type (Davis et al. 1968). Before this development, emphasis was laid on the production of constant shear-rate viscometers such as the Ferranti-Shirley cone-and-plate viscometer. This latter machine has a range of about 20 to 20,000 s$^{-1}$, whilst the Haake version has a range of about 1 to 1000 s$^{-1}$, in both cases a $10^3$-fold range. The Umstätter capillary viscometer, an earlier development, with a choice of capillaries, provides a $10^6$-fold range. Such instruments are suitable for the middle and upper regions of the general flow curve but they are not suitable for the resolution of the low shear-rate region. To do this, researchers used creep tests (§3.7.1) and devices like the plastometer (see, for example, Sherman, 1970, p 59), but there was no overlap between results from these instruments and those from the constant shear-rate devices. Hence the low shear-rate region could never be unequivocally linked with the high shear-rate region. This situation has now changed and the overlap has already been achieved for a number of materials.

Equations that predict the shape of the general flow curve need at least four parameters. One such is the Cross (1965) equation given by

$$\frac{\eta - \eta_\infty}{\eta_0 - \eta_\infty} = \frac{1}{\left(1 + (K\dot{\gamma})^m\right)}, \tag{2.2a}$$

or, what is equivalent,

$$\frac{\eta_0 - \eta}{\eta - \eta_\infty} = (K\dot{\gamma})^m, \tag{2.2b}$$

where $\eta_0$ and $\eta_\infty$ refer to the asymptotic values of viscosity at very low and very high shear rates respectively, $K$ is a constant parameter with the dimension of time and $m$ is a dimensionless constant.

A popular alternative to the Cross model is the model due to Carreau (1972)

$$\frac{\eta - \eta_\infty}{\eta_0 - \eta_\infty} = \frac{1}{\left(1 + (K_1\dot{\gamma})^2\right)^{m_1/2}}, \tag{2.3}$$

where $K_1$ and $m_1$ have a similar significance to the $K$ and $m$ of the Cross model.

By way of illustration, we give examples in Fig. 2.4 of the applicability of the Cross model to a number of selected materials.

It is informative to make certain approximations to the Cross model, because, in so doing, we can introduce a number of other popular and widely used viscosity models. * For example, for $\eta \ll \eta_0$ and $\eta \gg \eta_\infty$, the Cross model reduces to

$$\eta = \frac{\eta_0}{(K\dot{\gamma})^m}, \tag{2.4}$$

---

* We have used shear *rate* as the independent variable. However, we could equally well have employed the shear *stress* in this connection, with, for instance, the so-called Ellis model as the equivalent of the Cross model.

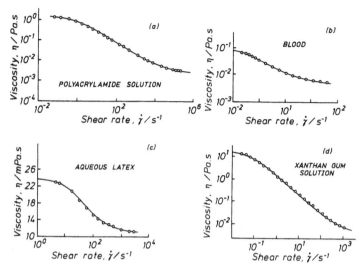

Fig. 2.4. Examples of the applicability of the Cross equation (eqn. (2.2a)): (a) 0.4% aqueous solution of polyacrylamide. Data from Boger (1977(b)). The solid line represents the Cross equation with $\eta_0 = 1.82$ Pa.s, $\eta_\infty = 2.6$ mPa.s, $K = 1.5$ s, and $m = 0.60$; (b) Blood (normal human, $Hb = 37\%$). Data from Mills et al. (1980). The solid line represents the Cross equation with $\eta_0 = 125$ mPa.s, $\eta_\infty = 5$ mPa.s, $K = 52.5$ s and $m = 0.715$; (c) Aqueous dispersion of polymer latex spheres. Data from Quemada (1978). The solid line represents the Cross equation with $\eta_0 = 24$ mPa.s, $\eta_\infty = 11$ mPa.s, $K = 0.018$ s and $m = 1.0$; (d) 0.35% aqueous solution of Xanthan gum. Data from Whitcomb and Macosko (1978). The solid line represents the Cross equation with $\eta_0 = 15$ Pa.s, $\eta_\infty = 5$ mPa.s, $K = 10$ s, $m = 0.80$.

which, with a simple redefinition of parameters can be written

$$\eta = K_2 \dot{\gamma}^{n-1}. \tag{2.5}$$

This is the well known '*power-law*' *model* and $n$ is called the power-law index. $K_2$ is called the 'consistency' (with the strange units of Pa.s$^n$).

Further, if $\eta \ll \eta_0$, we have

$$\eta = \eta_\infty + \frac{\eta_0}{(K\dot{\gamma})^m}, \tag{2.6}$$

which can be rewritten as

$$\eta = \eta_\infty + K_2 \dot{\gamma}^{n-1}. \tag{2.7}$$

This is called the *Sisko (1958) model*. If $n$ is set equal to zero in the Sisko model, we obtain

$$\eta = \eta_\infty + \frac{K_2}{\dot{\gamma}}, \tag{2.8a}$$

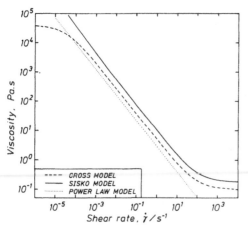

Fig. 2.5 Typical viscosity/shear rate graphs obtained using the Cross, power-law and Sisko models. Data for the Cross equation curve are the same as used in Fig. 2.3. The other curves represent the same data but have been shifted for clarity.

which, with a simple redefinition of parameters can be written

$$\sigma = \sigma_y + \eta_p \dot{\gamma}, \tag{2.8b}$$

where $\sigma_y$ is the yield stress and $\eta_p$ the plastic viscosity (both constant). This is the *Bingham model* equation.

The derived equations apply over limited parts of the flow curve. Figure 2.5 illustrates how the power-law fits only near the central region whilst the Sisko model fits in the mid-to-high shear-rate range.

The Bingham equation describes the shear stress/shear rate behaviour of many shear-thinning materials at low shear rates, but only over a one-decade range (approximately) of shear rate. Figures 2.6(a) and (b) show the Bingham plot for a synthetic latex, over two different shear-rate ranges. Although the curves fit the equation, the derived parameters depend on the shear-rate range. Hence, the use of the Bingham equation to characterize viscosity behaviour is unreliable in this case. However, the concept of yield stress is sometimes a very good approximation for practical purposes, such as in characterizing the ability of a grease to resist slumping in a roller bearing. Conditions under which this approximation is valid are that the local value of $n$ is small (say $< 0.2$) and the ratio $\eta_0/\eta_\infty$ is very large (say $> 10^9$).

The Bingham-type extrapolation of results obtained with a laboratory viscometer to give a yield stress has been used to predict the size of solid particles that could be permanently suspended in a gelled liquid. This procedure rarely works in practice for thickened aqueous systems because the liquid flows, albeit slowly, at stresses below this stress. The use of $\eta_0$ and Stokes' drag law gives a better prediction of the settling rate. Obviously, if this rate can be made sufficiently small the suspension becomes "non-settling" for practical purposes.

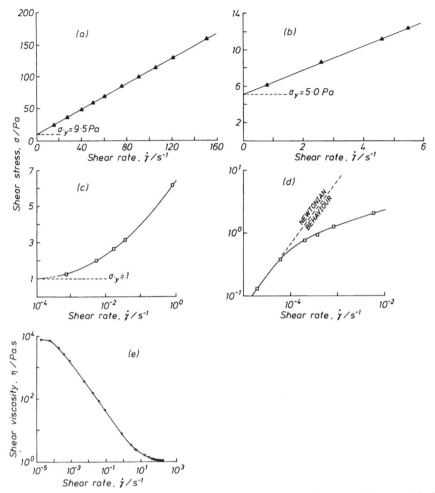

Fig. 2.6. Flow curves for a synthetic latex (taken from Barnes and Walters 1985): (a and b) Bingham plots over two different ranges of shear rate, showing two different intercepts; (c) Semi-logarithmic plot of data obtained at much lower shear rates, showing yet another intercept; (d) Logarithmic plot of data at the lowest obtainable shear rates, showing no yield-stress behaviour; (e) The whole of the experimental data plotted as viscosity versus shear rate on logarithmic scales.

The power-law model of eqn. (2.5) fits the experimental results for many materials over two or three decades of shear rate, making it more versatile than the Bingham model. It is used extensively to describe the non-Newtonian flow properties of liquids in theoretical analyses as well as in practical engineering applications. However, care should be taken in the use of the model when employed outside the range of the data used to define it. Table 2.3 contains typical values for the power-law parameters for a selection of well-known non-Newtonian materials.

The power-law model fails at high shear rates, where the viscosity must ulti-

TABLE 2.3
Typical power-law parameters of a selection of well-known materials for a particular range of shear rates.

| Material | $K_2 (\text{Pa.s}^n)$ | $n$ | Shear rate range $(\text{s}^{-1})$ |
|---|---|---|---|
| Ball-point pen ink | 10 | 0.85 | $10^0 - 10^3$ |
| Fabric conditioner | 10 | 0.6 | $10^0 - 10^2$ |
| Polymer melt | 10000 | 0.6 | $10^2 - 10^4$ |
| Molten chocolate | 50 | 0.5 | $10^{-1} - 10$ |
| Synovial fluid | 0.5 | 0.4 | $10^{-1} - 10^2$ |
| Toothpaste | 300 | 0.3 | $10^0 - 10^3$ |
| Skin cream | 250 | 0.1 | $10^0 - 10^2$ |
| Lubricating grease | 1000 | 0.1 | $10^{-1} - 10^2$ |

mately approach a constant value; in other words, the local value of $n$ must ultimately approach unity. This failure of the power-law model can be rectified by the use of the Sisko model, which was originally proposed for high shear-rate measurements on lubricating greases. Examples of the usefulness of the Sisko model in describing the flow properties of shear-thinning materials over four or five decades of shear rate are given in Fig. 2.7.

Attempts have been made to derive the various viscosity laws discussed in this

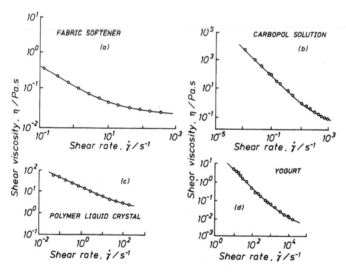

Fig. 2.7. Examples of the applicability of the Sisko model (eqn. (2.7)): (a) Commercial fabric softener. Data obtained by Barnes (unpublished). The solid line represents the Sisko model with $\eta_\infty = 24$ mPa.s, $K_2 = 0.11$ Pa.s $^n$ and $n = 0.4$; (b) 1% aqueous solution of Carbopol. Data obtained by Barnes (unpublished). The solid line represents the Sisko model with $\eta_\infty = 0.08$ Pa.s, $K_2 = 8.2$ Pa.s$^n$ and $n = 0.066$; (c) 40% Racemic poly-$\gamma$-benzyl glutamate polymer liquid crystal. Data points obtained from Onogi and Asada (1980). The solid line represents the Sisko model with $\eta_\infty = 1.25$ Pa.s, $K_2 = 15.5$ Pa.s$^n$, $n = 0.5$; (d) Commercial yogurt. Data points obtained from deKee et al. (1980). The solid line represents the Sisko model with $\eta_\infty = 4$ mPa.s, $K_2 = 34$ Pa.s$^n$ and $n = 0.1$.

section from microstructural considerations. However, these laws must be seen as being basically empirical in nature and arising from curve-fitting exercises.

### 2.3.3 The shear-thickening non-Newtonian liquid

It is possible that the very act of deforming a material can cause rearrangement of its microstructure such that the resistance to flow increases with shear rate. Typical examples of the shear-thickening phenomenon are given in Fig. 2.8. It will be observed that the shear-thickening region extends over only about a decade of shear rate. In this region, the power-law model can usually be fitted to the data with a value of $n$ greater than unity.

In almost all known cases of shear-thickening, there is a region of shear-thinning at lower shear rates.

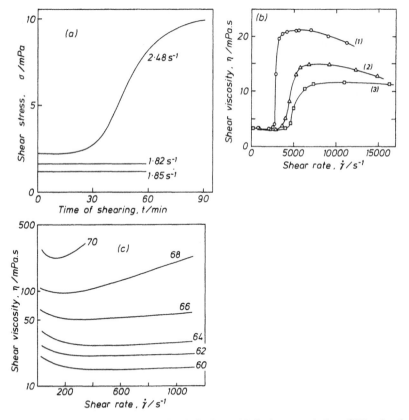

Fig. 2.8. Examples of shear-thickening behaviour: (a) Surfactant solution. CTA-sal. solution at 25° C, showing a time-effect (taken from Gravsholt 1979); (b) Polymer solution. Solution of anti-misting polymer in aircraft jet fuel, showing the effect of photodegradation during ($1$) 1 day, ($2$) 15 days, ($3$) 50 days exposure to daylight at room temperature (taken from Matthys and Sabersky 1987); (c) Aqueous suspensions of solid particles. Deflocculated clay slurries showing the effect of concentration of solids. The parameter is the %w/w concentration (taken from Beazley 1980).

### 2.3.4 Time effects in non-Newtonian liquids

We have so far assumed by implication that a given shear rate results in a corresponding shear stress, whose value does not change so long as the value of the shear rate is maintained. This is often not the case. The measured shear stress, and hence the viscosity, can either increase or decrease with time of shearing. Such changes can be reversible or irreversible.

According to the accepted definition, a gradual decrease of the viscosity under shear stress followed by a gradual recovery of structure when the stress is removed is called 'thixotropy'. The opposite type of behaviour, involving a gradual increase in viscosity, under stress, followed by recovery, is called 'negative thixotropy' or 'anti-thixotropy'. A useful review of the subject of time effects is provided by Mewis (1979).

Thixotropy usually occurs in circumstances where the liquid is shear-thinning (in the sense that viscosity levels decrease with increasing shear rate, other things being equal). In the same way, anti-thixotropy is usually associated with shear-thickening behaviour. The way that either phenomenon manifests itself depends on the type of test being undertaken. Figure 2.9 shows the behaviour to be expected from relatively inelastic colloidal materials in two kinds of test: the first involving step changes in applied shear rate or shear stress and the second being a loop test with the shear

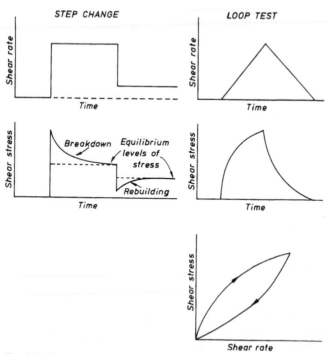

Fig. 2.9. Schematic representation of the response of an inelastic thixotropic material to two shear-rate histories.

rate increased continuously and linearly in time from zero to some maximum value and then decreased to zero in the same way.

If highly elastic colloidal liquids are subjected to such tests, the picture is more complicated, since there are contributions to the stress growth and decay from viscoelasticity.

The occurrence of thixotropy implies that the flow history must be taken into account when making predictions of flow behaviour. For instance, flow of a thixotropic material down a long pipe is complicated by the fact that the viscosity may change with distance down the pipe.

### 2.3.5 Temperature effects in two-phase non-Newtonian liquids

In the simplest case, the change of viscosity with temperature in two-phase liquids is merely a reflection of the change in viscosity of the continuous phase. Thus some aqueous systems at room temperature have the temperature sensitivity of water, i.e. 3% per °C. In other cases, however, the behaviour is more complicated. In dispersions, the suspended phase may go through a melting point. This will result in a sudden and larger-than-expected decrease of viscosity. In those dispersions, for which the viscosity levels arise largely from the temperature-sensitive colloidal interactions between the particles, the temperature coefficient will be different from that of the continuous phase. For detergent-based liquids, small changes in temperature can result in phase changes which may increase or decrease the viscosity dramatically.

In polymeric systems, the solubility of the polymer can increase or decrease with temperature, depending on the system. The coiled chain structure may become more open, resulting in an increase in resistance to flow. This is the basis of certain polymer-thickened multigrade oils designed to maintain good lubrication at high temperatures by partially offsetting the decrease in viscosity with temperature of the base oil (see also §6.11.2).

## 2.4 Viscometers for measuring shear viscosity

### 2.4.1 General considerations

Accuracy of measurement is an important issue in viscometry. In this connection, we note that it is possible in principle to calibrate an instrument in terms of speed, geometry and sensitivity. However, it is more usual to rely on the use of standardized Newtonian liquids (usually oils) of known viscosity. Variation of the molecular weight of the oils allows a wide range of viscosities to be covered. These oils are chemically stable and are not very volatile. They themselves are calibrated using glass capillary viscometers and these viscometers are, in turn, calibrated using the internationally accepted standard figure for the viscosity of water (1.002 mPa.s at 20.00°C, this value being uncertain to $\pm 0.25\%$). Bearing in mind the accumulated errors in either the direct or comparative measurements, the everyday measurement of viscosity must obviously be worse than the 0.25% mentioned above. In fact for mechanical instruments, accuracies of ten times this figure are more realistic.

| VISCOMETER TYPE | VARIABLES TO CHANGE | MEASUREMENT | CONVENIENCE | ROBUSTNESS |
|---|---|---|---|---|
| 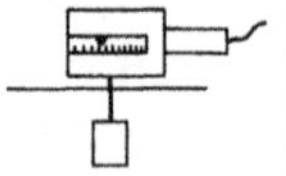 | BOB SPEED AND DIAMETER | COUPLE | * * | * |
| 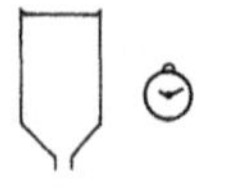 | EXIT TUBE DIAMETER | EFFLUX TIME | * * <br> * * | * * <br> * * |
| 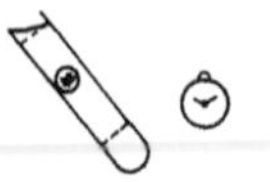 | BALL SIZE | VELOCITY | * * | * * |

Fig. 2.10. Examples of industrial viscometers with complicated flow fields, including star-ratings for convenience and robustness.

### 2.4.2 Industrial shop-floor instruments

Some viscometers used in industry have complicated flow and stress fields, although their operation is simple. In the case of Newtonian liquids, the use of such instruments does not present significant problems, since the instruments can be calibrated with a standard liquid. However, for non-Newtonian liquids, complicated theoretical derivations are required to produce viscosity information, and in some cases no amount of mathematical complication can generate consistent viscosity data (see, for example, Walters and Barnes 1980).

Three broad types of industrial viscometer can be identified (Fig. 2.10). The first type comprises rotational devices, such as the Brookfield viscometer. There is some hope of consistent interpretation of data from such instruments (cf. Williams 1979). The second type of instrument involves what we might loosely call "flow through constrictions" and is typified by the Ford-cup arrangement. Lastly, we have those that involve, in some sense, flow around obstructions such as in the Glen Creston falling-ball instrument (see, for example, van Wazer et al. 1963). Rising-bubble techniques can also be included in this third category.

For all the shop-floor viscometers, great care must be exercised in applying formulae designed for Newtonian liquids to the non-Newtonian case.

### 2.4.3 Rotational instruments; general comments

Many types of viscometer rely on rotational motion to achieve a simple shearing flow. For such instruments, the means of inducing the flow are two-fold: one can either drive one member and measure the resulting couple or else apply a couple and measure the subsequent rotation rate. Both methods were well established before the first World War, the former being introduced by Couette in 1888 and the latter by Searle in 1912.

There are two ways that the rotation can be applied and the couple measured: the first is to drive one member and measure the couple on the *same* member, whilst the other method is to drive one member and measure the couple on the *other*. In modern viscometers, the first method is employed in the Haake, Contraves, Fer-ranti-Shirley and Brookfield instruments; the second method is used in the Weissen-berg and Rheometrics rheogoniometers.

For couple-driven instruments, the couple is applied to one member and its rate of rotation is measured. In Searle's original design, the couple was applied with weights and pulleys. In modern developments, such as in the Deer constant-stress instrument, an electrical drag-cup motor is used to produce the couple. The couples that can be applied by the commercial constant stress instruments are in the range $10^{-6}$ to $10^{-2}$ Nm; the shear rates that can be measured are in the range $10^{-6}$ to $10^3$ s$^{-1}$, depending of course on the physical dimensions of the instruments and the viscosity of the material. The lowest shear rates in this range are equivalent to one complete revolution every two years; nevertheless it is often possible to take steady-state measurements in less than an hour.

As with all viscometers, it is important to check the calibration and zeroing from time to time using calibrated Newtonian oils, with viscosities within the range of those being measured.

### 2.4.4 The narrow-gap concentric-cylinder viscometer

If the gap between two concentric cylinders is small enough and the cylinders are in relative rotation, the test liquid enclosed in the gap experiences an almost constant shear rate. Specifically, if the radii of the outer and inner cylinders are $r_O$ and $r_1$, respectively, and the angular velocity of the inner is $\Omega_1$, (the other being stationary) the shear rate $\dot\gamma$ is given by

$$\dot\gamma = \frac{r_O \Omega_1}{r_O - r_1}. \tag{2.9}$$

For the gap to be classed as "narrow" and the above approximation to be valid to within a few percent, the ratio of $r_1$ to $r_O$ must be greater than 0.97.

If the couple on the cylinders is $C$, the shear stress in the liquid is given by

$$\sigma = \frac{C}{2\pi r_O^2 L}, \tag{2.10}$$

and from (2.9) and (2.10), we see that the viscosity is given by

$$\eta = \frac{C(r_O - r_1)}{2\pi r_O^3 \Omega_1 L}, \tag{2.11}$$

where $L$ is the effective immersed length of the liquid being sheared. This would be the real immersed length, $l$, if there were no end effects. However, end effects are likely to occur if due consideration is not given to the different shearing conditions which may exist in any liquid covering the ends of the cylinders.

One way to proceed is to carry out experiments at various immersed lengths, *l*, keeping the rotational rate constant. The extrapolation of a plot of *C* against *l* then gives the correction which must be added to the real immersed length to provide the value of the effective immersed length *L*. In practice, most commercial viscometer manufacturers arrange the dimensions of the cylinders such that the ratio of the depth of liquid to the gap between the cylinders is in excess of 100. Under these circumstances the end correction is negligible.

The interaction of one end of the cylinder with the bottom of the containing outer cylinder is often minimized by having a recess in the bottom of the inner cylinder so that air is entrapped when the viscometer is filled, prior to making measurements. Alternatively, the shape of the end of the cylinder can be chosen as a cone. In operation, the tip of the cone just touches the bottom of the outer cylinder container. The cone angle (equal to $\tan^{-1}\left[(r_O - r_1)/r_O\right]$) is such that the shear rate in the liquid trapped between the cone and the bottom is the same as that in the liquid between the cylinders. This arrangement is called the Mooney system, after its inventor.

### 2.4.5 The wide-gap concentric-cylinder viscometer

The limitations of very narrow gaps in the concentric-cylinder viscometer are associated with the problems of achieving parallel alignment and the difficulty of coping with suspensions containing large particles. For these reasons, in many commercial viscometers the ratio of the cylinder radii is less than that stated in §2.4.4; thus some manipulation of the data is necessary to produce the correct viscosity. This is a nontrivial operation and has been studied in detail by Krieger and Maron (1954). Progress can be readily made if it is assumed that the shear stress/shear rate relationship over the interval of shear rate in the gap can be described by the power-law model of eqn. (2.5). The shear rate in the liquid at the inner cylinder is then given by

$$\dot{\gamma} = \frac{2\Omega_1}{n\left(1 - b^{2/n}\right)}, \tag{2.12}$$

where *b* is the ratio of the inner to outer radius (i.e. $b = r_1/r_O$). Note that the shear rate is now dependent on the properties of the test liquid, unlike the narrow-gap instrument.

The shear stress in the liquid at the inner cylinder is given by

$$\sigma = \frac{C}{2\pi r_1^2 L}. \tag{2.13}$$

The value of *n* can be determined by plotting *C* versus $\Omega_1$ on a double-logarithmic basis and taking the slope at the value of $\Omega_1$ under consideration.

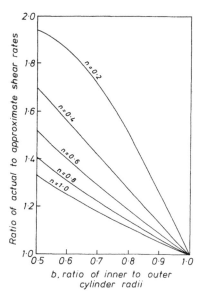

Fig. 2.11. Ratio of actual (eqn. 2.12) to approximate (eqn. 2.9) shear rates at the rotating cylinder as a function of the ratio of the inner to the outer cylinder radii, with $n$ the power law index as parameter.

The viscosity (measured at the inner-cylinder shear rate) is given by

$$\eta = \frac{Cn\left(1 - b^{2/n}\right)}{4\pi r_1^2 L\Omega_1}.$$                    (2.14)

The error involved in employing the narrow-gap approximation instead of the wide-gap expression, eqn. (2.14), is shown in Fig. 2.11. Clearly, using values of $b < 0.97$ gives unacceptable error when the liquid is shear-thinning ($n < 1$).

The lower limit of shear rate achievable in a rotational viscometer is obviously governed by the drive system. The upper limit, however, is usually controlled by the test liquid. One limit is the occurrence of viscous heating of such a degree that reliable correction cannot be made. However, there are other possible limitations. Depending on which of the cylinders is rotating, at a critical speed the simple circumferential streamline flow breaks down, either with the appearance of steady (Taylor) vortices or turbulence. Since both of these flows require more energy than streamline flow, the viscosity of the liquid *apparently* increases. In practical terms, for most commercial viscometers, it is advisable to consider the possibility of such disturbances occurring if the viscosity to be measured is less than about 10 mPa.s.

### 2.4.6 Cylinder rotating in a large volume of liquid

If we take the wide-gap Couette geometry to the extreme with the radius of the outer cylinder approaching infinity, the factor $(1 - b^{2/n})$ in (2.12) and (2.14)

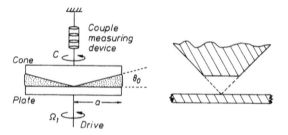

Fig. 2.12. The cone-and-plate viscometer. Cross-sectional diagram of one possible configuration, viz. cone on top, rotating plate and couple measured on the cone. The inset shows the form of truncation used in many instruments.

approaches unity. For a power-law liquid, the values of the shear rate and shear stress in the liquid at the rotating cylinder of radius $r_1$ are then given by

$$\dot{\gamma} = 2\Omega_1/n \tag{2.15}$$

and

$$\sigma = \frac{C}{2\pi r_1^2 L} . \tag{2.16}$$

Again, at any particular value of $\Omega_1$, $n$ can be calculated as the local value of $d(\ln C)/d(\ln \Omega_1)$. These equations are applicable to viscometers of the Brookfield type in which a rotating bob is immersed in a beaker of liquid. The technique is restricted to moderately low shear rates: $0.1 \text{ s}^{-1}$ to $10 \text{ s}^{-1}$ is a typical range.

### 2.4.7 The cone-and-plate viscometer

In the cone-and-plate geometry shown in Fig. 2.12, the shear rate is very nearly the same everywhere in the liquid provided the gap angle $\theta_0$ is small (see Chapter 4 for the details). The shear rate in the liquid is given by

$$\dot{\gamma} = \Omega_1/\theta_0, \tag{2.17}$$

where $\Omega_1$ is the angular velocity of the rotating platten. Note that the shear rate does not depend on the properties of the liquid.

The shear stress (measured via the couple $C$ on the cone) is given by

$$\sigma = \frac{3C}{2\pi a^3}, \tag{2.18}$$

where $a$ is the radius of the cone. Thus the viscosity is given by

$$\eta = \frac{3C\theta_0}{2\pi a^3 \Omega_1} . \tag{2.19}$$

If the liquid under investigation has a low viscosity, high rotational speeds are often necessary to produce torques large enough to be measured accurately. However, under these circumstances, 'secondary flows' may arise (see, for example, Walters 1975). The secondary flow absorbs extra energy, thus increasing the couple, which the unwary may mistakenly associate with shear-thickening. Cheng (1968) has provided an empirical formula which goes some way towards dealing with the problem.

All cone-and-plate instruments allow the cone to be moved away from the plate to facilitate sample changing. It is very important that the cone and plate be reset so that the tip of the cone lies in the surface of the plate. For a 1° gap angle and a cone radius of 50 mm, every 10 $\mu$m of error in the axial separation produces an additional 1% error in the shear rate.

To avoid error in contacting the cone tip (which might become worn) and the plate (which might become indented), the cone is often truncated by a small amount. In this case, it is necessary to set the virtual tip in the surface of the plate as shown in Fig. 2.12 (b). A truncated cone also facilitates tests on suspensions.

### 2.4.8 The parallel-plate viscometer

For torsional flow between parallel plates (see Fig. 2.13) the shear rate at the rim $(r = a)$ is given by

$$\dot{\gamma}_a = a\Omega_1/h. \tag{2.20}$$

It is this shear rate that finds its way into the interpretation of experimental data for torsional flow. It can be shown (Walters 1975, p. 52) that the viscosity is given by

$$\eta = \frac{3Ch}{2\pi a^4 \Omega_1}\left[1 + \frac{1}{3}\frac{d\ell nC}{d\ell n\Omega_1}\right] \tag{2.21}$$

where $C$ is the couple on one of the plates. For power-law models, eqn. (2.21) becomes

$$\eta = \frac{3Ch}{2\pi a^4 \Omega_1}\left[1 + \frac{n}{3}\right]. \tag{2.22}$$

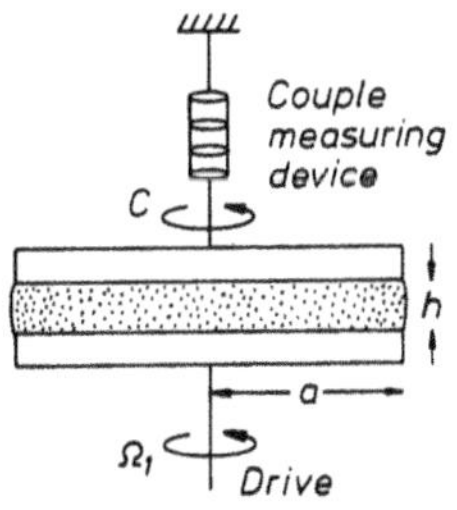

Fig. 2.13. Cross-sectional diagram of the torsional parallel-plate viscometer.

It will be noticed from eqn. (2.20) that the rim shear rate may be changed by adjusting either the speed $\Omega_1$ or the gap $h$.

In the torsional-balance rheometer, an adaptation of the parallel-plate viscometer (see Chapter 4 for the details), shear rates in the $10^4$ to $10^5$ s$^{-1}$ range have been attained.

### 2.4.9 Capillary viscometer

If a Newtonian liquid flows down a straight circular tube of radius $a$ at a volume flowrate $Q$ (see Fig. 2.14), the pressure gradient generated along it $(dP/dl)$ is given by the Poiseuille equation:

$$\frac{dP}{dl} = \frac{8Q\eta}{\pi a^4}. \tag{2.23}$$

In this situation, the shear stress in the liquid varies linearly from $(a/2)(dP/dl)$ at the capillary wall to zero at the centre line. For Newtonian liquids, the shear rate varies similarly from $4Q/(\pi a^3)$ in the immediate vicinity of the wall to zero at the centreline. If, however, the viscosity varies with shear rate the situation is more complex. Progress can be made by concentrating on flow near the wall. Analysis shows (cf. Walters 1975, Chapter 5) that for a non-Newtonian liquid, the shear rate at the wall is modified to

$$\dot{\gamma}_w = \frac{4Q}{\pi a^3}\left(\frac{3}{4} + \frac{1}{4}\frac{d\ln Q}{d\ln \sigma_w}\right), \tag{2.24}$$

whilst the shear stress at the wall $\sigma_w$, is unchanged at $(a/2)(dP/dl)$. The bracketed term in (2.24) is called the Rabinowitsch correction. Then finally

$$\eta(\dot{\gamma}_w) = \frac{\sigma_w}{\dot{\gamma}_w} = \frac{\pi a^4(dP/dl)}{8Q\left(\dfrac{3}{4} + \dfrac{1}{4}\dfrac{d\ln Q}{d\ln \sigma_w}\right)}. \tag{2.25}$$

When shear-thinning liquids are being tested, $d(\ln Q)/d(\ln \sigma_w)$ is greater than 1 and for power-law liquids is equal to $1/n$. Since $n$ can be as low as 0.2, the contribution

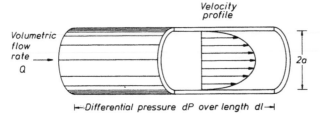

Fig. 2.14. Cutaway diagram of laminar Newtonian flow in a straight circular capillary tube.

of the $d(\ln Q)/d(\ln \sigma_w)$ factor to the bracketed term can be highly significant in determining the true wall shear rate.

Care has to be taken in defining and measuring the pressure gradient $dP/dl$. If the pressure in the external reservoir supplying the capillary and the receiving vessel are measured, then, unless the ratio of tube length to radius is very large ($> 100$), allowance must be made for entrance and exit effects. These arise from the following sources for all types of liquid:

*(i)* Viscous and inertial losses in the converging stream up to the entrance.

*(ii)* Redistribution of the entrance velocities to achieve the steady state velocity profile within the tube.

*(iii)* Similar effects to the above at the exit.

Formulae exist which account for these effects for Newtonian liquids, (*i*) and (*iii*) being associated with the names of Hagenbach and Couette (see, for example Kestin et al. 1973). However, these effects are small if the ratio of tube length to radius is 100 or more.

The main end effects can be avoided if at various points on the tube wall, well away from the ends, the pressures are measured by holes connected to absolute or differential pressure transducers. Any error arising from the flow of the liquid past the holes in the tube wall (see §4.4.1.II) is cancelled out when identical holes are used and the pressure gradient alone is required.

It is not often convenient to drill pressure-tappings, and a lengthy experimental programme may then be necessary to determine the type-(*i*) errors in terms of an equivalent pressure-drop and type-(*ii*) errors in terms of an extra length of tube. The experiments required can be deduced from the theoretical treatment of Kestin et al. (1973) and a recent application of them has been published by Galvin et al. (1981). If the liquid is highly elastic, an additional entrance and exit pressure drop arises from the elasticity. The so-called Bagley correction then allows an estimate of the elastic properties to be calculated. It is also used to provide an estimate of the extensional viscosity of the liquid (see §5.4.6).

Before leaving the discussion of the capillary viscometer, it is of interest to study the pressure-gradient/flow-rate relationship for the power-law model (2.5):

$$\frac{dP}{dl} = \frac{2K_2}{a} \left[ \frac{(3n+1)Q}{\pi na^3} \right]^n. \tag{2.26}$$

From this equation we see the effect of changes in such variables as pipe radius. For Newtonian liquids, the pressure drop for a given flow rate is proportional to the *fourth* power of the radius, but this is changed if the liquid is shear-thinning. For instance, if $n = \frac{1}{3}$, the pressure drop is proportional to the *square* of the radius. This is clearly important in any scale-up of pipe flow from pilot plant to factory operation.

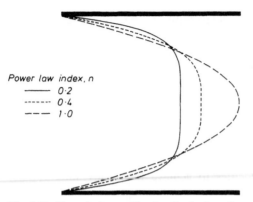

Fig. 2.15. The velocity profiles for the laminar flow of power-law liquids in a straight circular pipe, calculated for the same volumetric throughput. Note the increase in the wall shear rate and the increasingly plug-like nature of the flow as *n* decreases.

The velocity profile in pipe flow is parabolic for Newtonian liquids. For power-law liquids this is modified to

$$v(r) = \frac{Q(3n+1)}{\pi a^2 (n+1)} \left( 1 - \left( \frac{r}{a} \right)^{(n+1)/n} \right). \tag{2.27}$$

Figure 2.15 shows the effect of progressively decreasing the power-law index, i.e. increasing the degree of shear thinning. We see the increasing plug-like nature of the flow with, effectively, only a thin layer near the wall being sheared. This has important consequences in heat-transfer applications, where heating or cooling is applied to the liquid from the outside of the pipe. The overall heat transfer is partly controlled by the shear rate in the liquid near the pipe wall. For a power-law liquid, this shear rate is changed from the Newtonian value by a factor $[3 + (1/n)]/4$. This means that heat transfer is increased for shear-thinning liquids (n < 1) and decreased for shear-thickening liquids (n > 1), but the former is the larger effect.

### 2.4.10 Slit viscometer

Flow under an applied pressure gradient between two parallel stationary walls is known as slit flow. It is a two-dimensional analogue of capillary flow. The governing equations for slit flow are (cf. Walters 1975, Chapter 5)

$$\sigma_w = \frac{h}{2} \frac{dP}{dl} \tag{2.28}$$

and

$$\dot{\gamma}_w = \frac{2Q}{bh^2} \left( 2 + \frac{d \ln Q}{d \ln \sigma_w} \right), \tag{2.29}$$

where $h$ is the slit height and $b$ is the slit width, $Q$ is the flow rate and $dP/dl$ is the pressure gradient.

Slit flow forms the basis of the viscometer version of the Lodge stressmeter. The stressmeter is described more fully in Chapter 4. The viscometer version differs from that discussed in §4.4.3 in that the transducer $T_1$ in Fig. 4.13 is unnecessary and is replaced by a solid wall. The instrument has the advantage that shear rates in excess of $10^6$ s$^{-1}$ can be achieved with little interference from viscous heating.

### 2.4.11 On-line measurements

It is frequently necessary to monitor the viscosity of a liquid "on line" in a number of applications, particularly when the constitution or temperature of the liquid is likely to change. Of the viscometers described in this chapter, the capillary viscometer and the concentric-cylinder viscometer are those most conveniently adapted for such a purpose. For the former, for example, the capillary can be installed directly in series with the flow: the method has attractive features, but its successful application to non-Newtonian liquids is non-trivial.

Care must be taken with the on-line concentric-cylinder apparatus, since the interpretation of data from the resulting helical flow is not easy.

Other on-line methods involve obstacles in the flow channel: for example, a float in a vertical tapered tube, as in the Rotameter, will arrive at an equilibrium position in the tube depending on the precise geometry, the rate of flow, the viscosity and the weight of the obstacle. The parallel-plate viscometer has also been adapted for on-line measurement (see, for example, Noltingk 1975).

CHAPTER 3

# LINEAR VISCOELASTICITY

## 3.1 Introduction

The word 'viscoelastic' means the simultaneous existence of viscous and elastic properties in a material (cf. §1.2). It is not unreasonable to assume that all real materials are viscoelastic, i.e. in all materials, both viscous and elastic properties coexist. As was pointed out in the Introduction, the particular response of a sample in a given experiment depends on the time-scale of the experiment in relation to a natural time of the material. Thus, if the experiment is relatively slow, the sample will appear to be viscous rather than elastic, whereas, if the experiment is relatively fast, it will appear to be elastic rather than viscous. At intermediate time-scales mixed (viscoelastic) response is observed. The concept of a natural time of a material will be referred to again later in this chapter. However, a little more needs to be said about the assumption of viscoelasticity as a universal phenomenon. It is not a generally-held assumption and would be difficult to prove unequivocally. Nevertheless, experience has shown that it is preferable to assume that all real materials are viscoelastic rather than that some are not. Given this assumption, it is then incorrect to say that a liquid is Newtonian or that a solid is Hookean. On the other hand, it would be quite correct to say that such-and-such a material shows Newtonian, or Hookean, behaviour in a given situation. This leaves room for ascribing other types of behaviour to the material in other circumstances. However, most rheologists still refer to certain classes of liquid (rather than their behaviour) as being Newtonian and to certain classes of solid as being Hookean, even when they know that these materials can be made to deviate from the model behaviours. Indeed, it is done in this book! Old habits die hard. However, it is considered more important that an introductory text should point out that such inconsistencies exist in the literature rather than try to maintain a purist approach.

For many years, much labour has been expended in the determination of the linear viscoelastic response of materials. There are many reasons for this (see, for example, Walters 1975, p. 121, Bird et al. 1987(a), p. 225). First there is the possibility of elucidating the molecular structure of materials from their linear-viscoelastic response. Secondly, the material parameters and functions measured in the relevant experiments sometimes prove to be useful in the quality-control of industrial products. Thirdly, a background in linear viscoelasticity is helpful before proceeding to the much more difficult subject of *non*-linear viscoelasticity (cf. the relative simplicity of the mathematics in the present chapter with that in Chapter 8 which essentially deals with non-linear viscoelasticity). Finally, a further motivation for some past studies of viscoelasticity came from tribology, where knowledge of the steady shear viscosity function $\eta(\dot{\gamma})$ discussed in §2.3 was needed at high shear rates ($10^6$ s$^{-1}$ or higher). Measurements of this function on low-viscosity "Newtonian" lubricants at high shear rates were made difficult by such factors as viscous heating, and this led to a search for an analogy between shear viscosity and the corresponding dynamic viscosity determined under linear viscoelastic conditions, the argument being that the latter viscosity was easier to measure (see, for example, Dyson 1970).

Many books on rheology and rheometry have sections on linear viscoelasticity. We recommend the text by Ferry (1980) which contains a wealth of information and an extensive list of references. Mathematical aspects of the subject are also well covered by Gross (1953) and Staverman and Schwarzl (1956).

## 3.2 The meaning and consequences of linearity

The development of the mathematical theory of linear viscoelasticity is based on a "superposition principle". This implies that the response (e.g. strain) at any time is directly proportional to the value of the initiating signal (e.g. stress). So, for example, doubling the stress will double the strain. In the linear theory of viscoelasticity, the differential equations are linear. Also, the coefficients of the time differentials are constant. These constants are material parameters, such as viscosity coefficient and rigidity modulus, and they are not allowed to change with changes in variables such as strain or strain rate. Further, the time derivatives are ordinary partial derivatives. This restriction has the consequence that the linear theory is applicable only to small changes in the variables.

We can now write down a general differential equation for linear viscoelasticity as follows:

$$
\left(1 + \alpha_1 \frac{\partial}{\partial t} + \alpha_2 \frac{\partial^2}{\partial t^2} + \ldots + \alpha_n \frac{\partial^n}{\partial t^n}\right)\sigma
$$

$$
= \left(\beta_0 + \beta_1 \frac{\partial}{\partial t} + \beta_2 \frac{\partial^2}{\partial t^2} + \ldots + \beta_m \frac{\partial^m}{\partial t^m}\right)\gamma, \tag{3.1}
$$

where $n = m$ or $n = m - 1$ (see for example, Oldroyd 1964). Note that for simplicity we have written (3.1) in terms of the shear stress $\sigma$ and the strain $\gamma$, relevant to a simple shear of the sort discussed in Chapter 1, except that we now allow $\sigma$ and $\gamma$ to be functions of the time $t$. However, we emphasise that other types of deformation could be included without difficulty, with the stress and strain referring to that particular deformation process. Mathematically, this means that we could replace the scalar variables $\sigma$ and $\gamma$ by their tensor generalizations. For example, $\sigma$ could be replaced by the stress tensor $\sigma_{ij}$.

## 3.3 The Kelvin and Maxwell models

We now consider some important special cases of eqn. (3.1). If $\beta_0$ is the only non-zero parameter, we have

$$\sigma = \beta_0 \gamma, \tag{3.2}$$

which is the equation of Hookean elasticity (i.e. linear solid behaviour) with $\beta_0$ as the rigidity modulus. If $\beta_1$ is the only non-zero parameter, we have

$$\sigma = \beta_1 \frac{\partial \gamma}{\partial t}, \tag{3.3}$$

or

$$\sigma = \beta_1 \dot{\gamma} \tag{3.4}$$

in our notation. This represents Newtonian viscous flow, the constant $\beta_1$ being the coefficient of viscosity.

If $\beta_0$ ($= G$) and $\beta_1$ ($= \eta$) are both non-zero, whilst the other constants are zero, we have

$$\sigma = G\gamma + \eta \dot{\gamma}, \tag{3.5}$$

which is one of the simplest models of viscoelasticity. It is called the 'Kelvin model', although the name 'Voigt' is also used. If a stress $\bar{\sigma}$ is suddenly applied at $t = 0$ and held constant thereafter, it is easy to show that, for the Kelvin model,

$$\gamma = (\bar{\sigma}/G)[1 - \exp(-t/\tau_K)], \tag{3.6}$$

where $\tau_K$ has been written for the ratio $\eta/G$. It has the dimension of time and controls the rate of growth of strain $\gamma$ following the imposition of the stress $\bar{\sigma}$. Figure 3.1 shows the development of the dimensionless group $\gamma G/\bar{\sigma}$ diagrammatically. The equilibrium value of $\gamma$ is $\bar{\sigma}/G$; hence $\gamma G/\bar{\sigma} = 1$, which is also the value for the Hooke model. The difference between the two models is that, whereas the

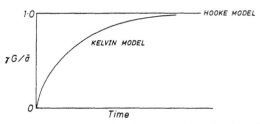

Fig. 3.1 Growth of strain $\gamma$ following the application of stress $\bar{\sigma}$ at time $t = 0$ for a Kelvin model and Hooke model.

Hooke model reaches its final value of strain "instantaneously", in the Kelvin model the strain is retarded. The time constant $\tau_K$ is accordingly called the 'retardation time'. The word instantaneously is put in quotation marks because in practice the strain could not possibly grow in zero time even in a perfectly elastic solid, because the stress wave travels at the speed of sound, thus giving rise to a delay.

It is useful at this stage to introduce "mechanical models", which provide a popular method of describing linear viscoelastic behaviour. These one-dimensional mechanical models consist of springs and dashpots so arranged, in parallel or in series, that the overall system behaves analogously to a real material, although the elements themselves may have no direct analogues in the actual material. The correspondence between the behaviour of a model and a real material can be achieved if the differential equation relating force, extension and time for the model is the same as that relating stress, strain and time for the material, i.e. this method is equivalent to writing down a differential equation relating stress and strain, but it has a practical advantage in that the main features of the behaviour of a material can often be inferred by inspection of the appropriate model, without going into the mathematics in detail.

In mechanical models, Hookean deformation is represented by a spring (i.e. an element in which the force is porportional to the extension) and Newtonian flow by a dashpot (i.e. an element in which the force is porportional to the rate of extension) as shown in Fig. 3.2. The analogous rheological equations for the spring and the dashpot are (3.2) (with $\beta_0 = G$) and (3.4) (with $\beta_1 = \eta$), respectively. The behaviour of more complicated materials is described by connecting the basic elements in series or in parallel.

The Kelvin model results from a parallel combination of a spring and a dashpot (Fig. 3.3(a)). A requirement on the interpretation of this and all similar diagrams is that the horizontal connectors remain parallel at all times. Hence the extension (strain) in the spring is at all times equal to the extension (strain) in the dashpot. Then it is possible to set up a balance equation for the forces (stresses) acting on a connector. The last step is to write the resulting equation in terms of stresses and strains. Hence, for the Kelvin model the total stress $\sigma$ is equal to the sum of the stresses in each element. Therefore

$$\sigma = \sigma_E + \sigma_V \tag{3.7}$$

(a)                                    (b)

Fig. 3.2 Diagrammatic representations of ideal rheological behaviour: (a) The Hookean spring; (b) The Newtonian dashpot.

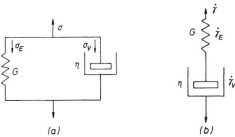

(a)                                              (b)

Fig. 3.3 The simplest linear viscoelastic models: (a) The Kelvin model; (b) The Maxwell model.

in the obvious notation, and using eqns. (3.2) and (3.4) (with $\beta_0 = G$ and $\beta_1 = \eta$) we have

$$\sigma = G\gamma + \eta\dot{\gamma}. \tag{3.8}$$

This is identical to eqn. (3.5), which was a very simple case of the general linear differential equation (3.1). It is readily seen from the diagram of the Kelvin model that after sudden imposition of a shear stress $\bar{\sigma}$, the spring will eventually reach the strain given by $\bar{\sigma}/G$, but that the dashpot will retard the growth of the strain and, the higher the viscosity, the slower will be the response.

Another very simple model is the so-called 'Maxwell model' *. The differential equation for the model is obtained by making $\alpha_1$ and $\beta_1$ the only non-zero material parameters, so that

$$\sigma + \tau_M\dot{\sigma} = \eta\dot{\gamma}, \tag{3.9}$$

where we have written $\alpha_1 = \tau_M$ and $\beta_1 = \eta$.

If a particular strain rate $\bar{\gamma}$ is suddenly applied at $t = 0$ and held at that value for subsequent times, we can show that, for $t > 0$,

$$\sigma = \eta\bar{\gamma}\left[1 - \exp(-t/\tau_M)\right], \tag{3.10}$$

which implies that on start-up of shear, the stress growth is delayed; the time constant in this case is $\tau_M$. On the other hand, if a strain rate which has had a constant value $\bar{\gamma}$ for $t < 0$ is suddenly removed at $t = 0$, we can show that, for $t \geqslant 0$,

$$\sigma = \eta\bar{\gamma}\exp(-t/\tau_M). \tag{3.11}$$

Hence the stress relaxes exponentially from its equilibrium value to zero (see Fig. 3.4). The rate constant $\tau_M$ is called the 'relaxation time'.

---

* Recall the discussion in §1.2 concerning the influence of J.C. Maxwell on the introduction of the concept of viscoelasticity in a fluid.

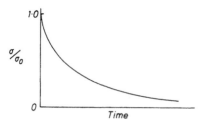

Fig. 3.4 Decay of stress $\sigma$ following the cessation of steady shear at time $t = 0$ for a Maxwell model, where $\sigma_0 = \eta\bar{\gamma}$.

The pictorial Maxwell model is a spring connected in series with a dashpot (see Fig. 3.3(b)). In this case, the strains, or equally strain-rates, are additive; hence the total rate of shear $\dot{\gamma}$ is the sum of the rates of shear of the two elements. Thus

$$\dot{\gamma} = \dot{\gamma}_E + \dot{\gamma}_V, \tag{3.12}$$

which leads to

$$\dot{\gamma} = \frac{\dot{\sigma}}{G} + \frac{\sigma}{\eta} \tag{3.13}$$

or, after rearrangement,

$$\sigma + \tau_M\dot{\sigma} = \eta\dot{\gamma}, \tag{3.14}$$

in which $\tau_M$ has been written for $\eta/G$. This equation is the same as eqn. (3.9) which arose as a special case of the general differential equation.

The next level of complexity in the linear viscoelastic scheme is to make three of the material parameters of eqn. (3.1) non zero. If $\alpha_1$, $\beta_1$ and $\beta_2$ are taken to be non-zero we have the "Jeffreys model". In the present notation, the equation is

$$\sigma + \tau_M\dot{\sigma} = \eta(\dot{\gamma} + \tau_J\ddot{\gamma}), \tag{3.15}$$

which has two time constants $\tau_M$ and $\tau_J$. With a suitable choice of the three model parameters it is possible to construct two alternative spring–dashpot models which correspond to the same mechanical behaviour as eqn. (3.15). One is a simple extension of the Kelvin model and the other a simple extension of the Maxwell model as shown in Fig. 3.5.

We note with interest that an equation of the form (3.15) was derived mathematically by Fröhlich and Sack (1946) for a dilute suspension of solid elastic spheres in a viscous liquid, and by Oldroyd (1953) for a dilute emulsion of one incompressible viscous liquid in another. When the effect of interfacial slipping is taken into account in the *dilute* suspension case, Oldroyd (1953) showed that two further non-zero parameters ($\alpha_2$ and $\beta_2$) are involved.

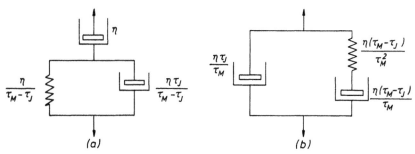

Fig. 3.5 Spring–dashpot equivalents of the Jeffreys model. The values of the constants of the elements are given in terms of the three material parameters of the model (eqn. 3.15).

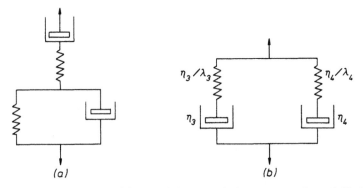

Fig. 3.6 The Burgers model: (a) and (b) are equivalent representations of this 4-parameter linear model.

Finally, in this preliminary discussion of the successive build-up of model complexity, we draw attention to the so-called "Burgers model". This involves four simple elements and takes the mechanically-equivalent forms shown in Fig. 3.6.

In terms of the parameters of the Maxwell-type representation (Fig. 3.6(b)), the associated constitutive equation for the Burgers model has the form

$$\sigma + (\lambda_3 + \lambda_4)\dot{\sigma} + \lambda_3\lambda_4\ddot{\sigma} = (\eta_3 + \eta_4)\dot{\gamma} + (\lambda_4\eta_3 + \lambda_3\eta_4)\ddot{\gamma}. \tag{3.16}$$

In this equation the $\lambda$s are time constants, the symbol $\lambda$ being almost as common as $\tau$ in the rheological litrature.

## 3.4 The relaxation spectrum

It is certainly possible to envisage more complicated models than those already introduced, but Roscoe (1950) showed that all models, irrespective of their complexity, can be reduced to two canonical forms. These are usually taken to be the generalized Kelvin model and the generalized Maxwell model (Fig. 3.7). The generalized Maxwell model may have a finite number or an enumerable infinity of Maxwell elements, each with a different relaxation time.

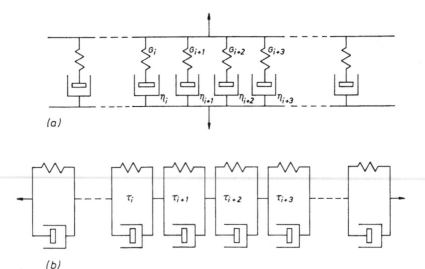

(a)

(b)

Fig. 3.7 Canonical spring–dashpot models: (a) Distribution of Maxwell relaxation processes; (b) Distribution of Kelvin retardation processes.

By a suitable choice of the model parameters, the canonical forms themselves can be shown to be mechanically equivalent and Alfrey (1945) has given methods for computing the parameters of one canonical form from those of the other. In the same paper, Alfrey also showed how a linear differential equation can be obtained for either of the canonical forms and vice versa. In other words, the three methods of representing viscoelastic behaviour (the differential equation (3.1) and the two canonical forms of mechanical model of Fig. 3.7 are equivalent and one is free to choose any one of them as a basis for generalization to materials requiring a continuous infinity of parameters to specify them.

In order to generalize from an enumerable infinity to a continuous distribution of relaxation times, we choose to start with the simple Maxwell model, whose be-haviour is characterized by the differential equation (3.9) or what is equivalent

$$\sigma(t) = \frac{\eta}{\tau} \int_{-\infty}^{t} \exp[-(t-t')/\tau] \dot{\gamma}(t') \, dt', \tag{3.17}$$

where we have dropped the subscript M in $\tau_M$ to enable us to generalize eqn. (3.17) without introducing a clumsy notation. *

Considering next, a number, $n$, of discrete Maxwell elements connected in parallel as in Fig. 3.7(a), we can generalize eqn. (3.17), with the aid of the

---

* The integral equation (3.17) is obtained by solving the differential equation (3.9) by standard techniques.

superposition principle, to give

$$\sigma(t) = \sum_{i=1}^{n} \frac{\eta_i}{\tau_i} \int_{-\infty}^{t} \exp\left[-(t-t')/\tau_i\right] \dot\gamma(t') \, dt', \tag{3.18}$$

where $\eta_i$ and $\tau_i$ now correspond to the $i$th Maxwell element.

The theoretical extension to a continuous distribution of relaxation times can be carried out in a number of ways. For example, we may proceed as follows.

The "distribution function of relaxation times" (or "relaxation spectrum") $N(\tau)$ may be defined such that $N(\tau) \, d\tau$ represents the contributions to the total *viscosity* of all the Maxwell elements with relaxation times lying between $\tau$ and $\tau + d\tau$. The relevant equation then becomes (on generalizing (3.18))

$$\sigma(t) = \int_0^\infty \frac{N(\tau)}{\tau} \int_{-\infty}^{t} \exp\left[-(t-t')/\tau\right] \dot\gamma(t') \, dt' \, d\tau, \tag{3.19}$$

and if we introduce the "relaxation function" $\phi$, defined by

$$\phi(t-t') = \int_0^\infty \frac{N(\tau)}{\tau} \exp\left[-(t-t')/\tau\right] d\tau, \tag{3.20}$$

eqn. (3.19) becomes

$$\sigma(t) = \int_{-\infty}^{t} \phi(t-t') \dot\gamma(t') \, dt'. \tag{3.21}$$

We remark that we could have immediately written down an equation like (3.21) on the basis of Boltzmann's superposition principle.

It is also possible to proceed from eqn. (3.18) by introducing a distribution function $H(\tau)$ such that $H(\tau) \, d\tau$ represents the contribution to the *elasticity modulus* of the processes with relaxation times lying in the interval $\tau$ and $\tau + d\tau$. Further, other workers have used a spectrum of relaxation frequencies $\overline{H}(\log F)$ where $F = 1/(2\pi\tau)$. The relationships between these functions are

$$(N(\tau)/\tau) \, d\tau = H(\tau) \, d\tau = \overline{H}(\log F) \, d(\log F). \tag{3.22}$$

In a *slow steady* motion which has been in existence indefinitely (i.e. $\dot\gamma$ is small, and independent of time) eqn. (3.21) reduces to

$$\sigma = \eta_0 \dot\gamma, \tag{3.23}$$

where

$$\eta_0 = \int_{-\infty}^{t} \phi(t-t') \, dt' = \int_0^\infty \phi(\xi) \, d\xi,$$

in which $\xi$ has been written for the time interval $(t - t')$. The variable $\xi$ is the one which represents the time-scale of the rheological history. It is also easy to show from eqns. (3.19), (3.21) and (3.22) that

$$\eta_0 = \int_0^\infty N(\tau)\, d\tau = \int_0^\infty \tau H(\tau)\, d\tau = \int_{-\infty}^\infty \frac{\overline{H}(\log F)}{2\pi F}\, d(\log F). \tag{3.24}$$

We see from eqn. (3.23) that $\eta_0$ can be identified with the limiting viscosity at small rates of shear, as observed in steady state experiments. Thus, the equations in (3.24) provide useful normalization conditions on the various relaxation spectra. It is also of interest to note that $\eta_0$ is equal to the area under the $N(\tau)$ spectrum, whilst it is equal to the first moment of the $H(\tau)$ spectrum.

### 3.5 Oscillatory shear

It is instructive to discuss the response of viscoelastic materials to a small-amplitude oscillatory shear, since this is a popular deformation mode for investigating linear viscoelastic behaviour.

Let

$$\gamma(t') = \gamma_0 \exp(i\omega t'), \tag{3.25}$$

where $i = \sqrt{-1}$, $\omega$ is the frequency and $\gamma_0$ is a strain amplitude which is small enough for the linearity constraint to be satisfied. The corresponding strain rate is given by

$$\dot{\gamma}(t') = i\omega\gamma_0 \exp(i\omega t'),$$

and, if this is substituted into the general integral equation (3.21), we obtain

$$\sigma(t) = i\omega\gamma_0 \exp(i\omega t) \int_0^\infty \phi(\xi) \exp(-i\omega\xi)\, d\xi. \tag{3.26}$$

In oscillatory shear we define a 'complex shear modulus' $G^*$, through the equation

$$\sigma(t) = G^*(\omega)\gamma(t) \tag{3.27}$$

and, from eqns. (3.25), (3.26) and (3.27), we see that

$$G^*(\omega) = i\omega \int_0^\infty \phi(\xi) \exp(-i\omega\xi)\, d\xi. \tag{3.28}$$

It is customary to write

$$G^* = G' + iG'' \tag{3.29}$$

and $G'$ and $G''$ are referred to as the 'storage modulus' and 'loss modulus', respectively. $G'$ is also called the dynamic rigidity. If we now consider, for the purpose of illustration, the special case of the Maxwell model given by eqn. (3.9) or eqn. (3.14) (with $\tau_M = \tau$) we can show that

$$G^* = \frac{i\omega\eta}{1 + i\omega\tau}, \quad \text{or alternatively} \quad G^* = \frac{i\omega\tau G}{1 + i\omega\tau}, \tag{3.30}$$

and

$$G' = \frac{\eta\tau\omega^2}{1 + \omega^2\tau^2}, \quad \text{or alternatively} \quad G' = \frac{G\omega^2\tau^2}{1 + \omega^2\tau^2}, \tag{3.31}$$

$$G'' = \frac{\eta\omega}{1 + \omega^2\tau^2}, \quad \text{or alternatively} \quad G'' = \frac{G\omega\tau}{1 + \omega^2\tau^2}. \tag{3.32}$$

To some readers, the use of the complex quantity $\exp(i\omega t)$ to represent oscillatory motion may be unfamiliar. The alternative procedure is to use the more obvious wave-forms represented by the sine and cosine functions, and we now illustrate the procedure for the simple Maxwell model.

Let

$$\gamma = \gamma_0 \cos\omega t. \tag{3.33}$$

Thus, the strain rate is

$$\dot{\gamma} = -\gamma_0\omega \sin\omega t, \tag{3.34}$$

and if this is substituted into the equation for the Maxwell model, a first order linear differential equation is obtained, with solution

$$\sigma = \frac{\eta\omega\gamma_0}{(1 + \omega^2\tau^2)}(\omega\tau \cos\omega t - \sin\omega t). \tag{3.35}$$

The part of the stress *in phase* with the applied strain is obtained by putting $\sin\omega t$ equal to zero and is written $G'\gamma$. The part of the stress which is *out of phase* with the applied strain is obtained by setting $\cos\omega t$ equal to zero and is written $(G''/\omega)\dot{\gamma}$. Hence

$$G' = \frac{\eta\tau\omega^2}{1 + \omega^2\tau^2}, \tag{3.36}$$

$$G'' = \frac{\eta\omega}{1 + \omega^2\tau^2}, \tag{3.37}$$

in agreement with (3.31) and (3.32) as expected.

Returning now to the more convenient complex representation of the oscillatory motion, we remark that as an alternative to the complex shear modulus, we can define 'complex viscosity' $\eta^*$, as the ratio of the shear stress $\sigma$ to the rate of shear $\dot{\gamma}$. Thus

$$\sigma(t) = \eta^* \ \dot{\gamma}(t), \tag{3.38}$$

and it follows that, for the general integral representation,

$$\eta^*(\omega) = \int_0^\infty \phi(\xi) \exp(-i\omega\xi) \ d\xi. \tag{3.39}$$

We now write

$$\eta^* = \eta' - i\eta'', \tag{3.40}$$

noting that $\eta'$ is usually given the name 'dynamic viscosity'. The parameter $\eta''$ has no special name but it is related to the dynamic rigidity through $G' = \eta''\omega$. It is also easy to deduce that $G'' = \eta'\omega$.

It is conventional to plot results of oscillatory tests in terms of the dynamic viscosity $\eta'$ and the dynamic rigidity $G'$. Figure 3.8 shows plots of the normalized dynamic functions of the Maxwell model as functions of $\omega\tau$, the normalized, or reduced, frequency. The notable features are the considerable fall in normalized $\eta'$ and the comparable rise in normalized $G'$ which occur together over a relatively narrow range of frequency centred on $\omega\tau = 1$. The changes in these functions are virtually complete in two decades of frequency. These two decades mark the viscoelastic zone. Also, in the many decades of frequency that are, in principle, accessible on the low frequency side of the relaxation region, the model displays a viscous response ($G' \sim 0$). In contrast, at high frequencies, the response is elastic ($\eta' \sim 0$). From Fig. 3.8, the significance of $\tau$ as a characteristic natural time for the Maxwell model is clear.

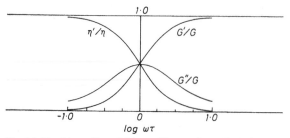

Fig. 3.8 The Maxwell model in oscillatory shear. Variation of the normalized moduli and viscosity with normalized frequency ($\tau = \eta/G$).

In the literature, other methods of characterizing linear viscoelastic behaviour are to be found. For example, it is possible to define a 'complex shear compliance' $J^*$. By definition

$$\gamma(t) = J^*(\omega)\sigma(t) \tag{3.41}$$

in an oscillatory shear, with

$$J^* = J' - iJ''. \tag{3.42}$$

It is important to note that, although $J^* = 1/G^*$, the components are not similarly related. Thus $J' \neq 1/G'$ and $J'' \neq 1/G''$.

The second alternative method of characterizing linear viscoelastic response is to plot $G'$ and the 'loss angle' $\delta$. In this method, it is assumed that for an applied oscillatory strain given by eqn. (3.25), the stress will have a similar form, but its phase will be in advance of the strain by an angle $\delta$. Then,

$$\sigma(t) = \sigma_0 \exp[i(\omega t + \delta)]. \tag{3.43}$$

It is not difficult to show that

$$\tan \delta = G''/G'. \tag{3.44}$$

Figure 3.9 shows how $\delta$ and $G'/G$ (where $G = \tau\eta$) vary with the normalized frequency for the Maxwell model. At high values of the frequency, the response, as already noted, is that of an elastic solid. In these conditions the stress is in phase with the applied strain. On the other hand, at very low frequencies, where the response is that of a viscous liquid, the stress is 90° ahead of the strain. The change from elastic to viscous behaviour takes place over about two decades of frequency. This latter observation has already been noted in connection with Fig. 3.8. In Fig. 3.10, we show the wave-forms for the oscillatory inputs and outputs. Figure 3.10(a) represents an experiment in the viscoelastic region. Figure 3.10(b) represents very high and very low frequency behaviour where the angle $\delta$ is 0° or 90°, respectively.

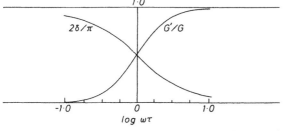

Fig. 3.9 The Maxwell model in oscillatory shear. Variation of the normalized storage modulus and phase angle with normalized frequency.

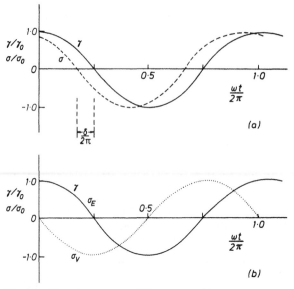

Fig. 3.10 Wave-forms for oscillatory strain input and stress output: (a) Solid line (———) strain according to eqns. (3.25) and (3.33); dashed line (- - - - - -) stress in advance by angle $\delta$, according to eqn. (3.43); (b) Solid line (———) strain input and also stress output for elastic behaviour; dotted line ($\cdots\cdots$) stress output for viscous behaviour.

Note that although the stress is $90\,°$ in advance of the shear strain for the viscous liquid, it is in phase with the rate of shear.

## 3.6 Relationships between functions of linear viscoelasticity

In previous sections we have introduced a number of different functions which can all be used to characterize linear viscoelastic behaviour. These range from complex moduli to relaxation function and spectra. They are not independent, of course, and we have already given mathematical relationships between some of the functions. For example, eqn. (3.28), which is fairly typical of the complexities involved, relates the complex shear modulus $G^*$ to the relaxation function $\phi$. Equation (3.28) is an integral transform and the determination of $\phi$ from $G^*$ can be accomplished by inverting the transform. There is nothing sophisticated, therefore, in determining one viscoelastic function from another: although this is a statement "in principle", and much work has been carried out on the non-trivial problem of inverting transforms when experimental data are available only over a limited range of the variables (like frequency of oscillation). The general problem of determining one viscoelastic function from another was discussed in detail by Gross (1953) and practical methods are dealt with by, for example, Schwarzl and Struik (1967).

Nowadays most experimental data from linear viscoelasticity experiments are presented in the form of graphs of components of the dynamic parameters (such as complex modulus) and are rarely transformed into the relaxation function or the relaxation spectrum.


## 3.7 Methods of measurement

Two different types of method are available to determine linear viscoelastic behaviour: *static* and *dynamic*. Static tests involve the imposition of a step change in stress (or strain) and the observation of the subsequent development in time of the strain (or stress). Dynamic tests involve the application of a harmonically varying strain.

In this section we shall be concerned with the main methods in the above classification. Attention will be focussed on the principles of the selected methods and none will be described in detail. The interested reader is referred to the detailed texts of Walters (1975) and Whorlow (1980) for further information.

An important point to remember is that, whatever the method adopted, the experimenter must check that measurements are made in the linear range; otherwise the results will be dependent on experimental details and will not be unique to the material. The test for linearity is to check that the computed viscoelastic functions are independent of the magnitude of the stresses and strains applied.

### 3.7.1 Static methods

The static methods are either 'creep' tests at constant stress or relaxation tests at constant strain (see Figs. 3.11 and 3.12). In theory, the input stress or strain, whether it is an increase or a decrease, is considered to be applied instantaneously. This cannot be true in practice, because of inertia in the loading and measuring systems and the delay in transmitting the signal across the test sample, determined by the speed of sound. As a general rule, the time required for the input signal to reach its steady value must be short compared to the time over which the ultimate varying output is to be recorded. This usually limits the methods to materials which have relaxation times of at least a few seconds. A technique for estimating whether apparatus inertia is influencing results is to deliberately change the inertia, by

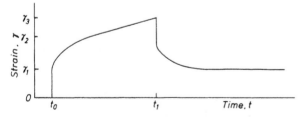

Fig. 3.11 Typical creep curve of strain $\gamma$ plotted against time $t$. A constant stress was applied at $t = t_0$ and removed at $t = t_1$. The strain comprises three regions: instantaneous (0 to $\gamma_1$); retardation ($\gamma_0$ to $\gamma_2$); constant rate ($\gamma_2$ to $\gamma_3$). In linear behaviour the instantaneous strains on loading and unloading are equal and the ratio of stress to instantaneous strain is independent of stress; the constant-rate strain ($\gamma_2$ to $\gamma_3$) is not recovered.

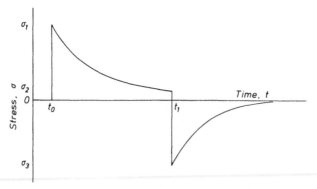

Fig. 3.12 Typical relaxation curve of stress $\sigma$ plotted against time $t$. A constant strain was applied at $t = t_0$ and reversed at $t = t_1$. In linear behaviour the instantaneous changes of stress from 0 to $\sigma_1$ and $\sigma_2$ to $\sigma_3$ are equal and the ratio of instantaneous stress to strain is independent of strain. The incomplete relaxation at $t = t_1$ may indicate either that further relaxation would occur in a longer time, or, that the material at very low deformation behaves like a Hookean solid and a residual stress would persist indefinitely.

adding weights for example, and checking the effect on the derived viscoelastic functions.

The basic apparatus for static tests is simple. Once the shape and means of holding the specimen have been decided upon, it is necessary to apply the input signal and measure, and record, the output. It is easier to measure strain, or deformation, than stress. Hence, creep tests have been much more common than relaxation tests.

The geometry chosen for static tests depends largely on the material to be tested. For solid-like materials, it is usually not difficult to fashion a long slender specimen for a tensile or torsional experiment. Liquid-like material can be tested in simple shear with the concentric-cylinder and cone- and-plate geometries and constant-stress rheometers are commercially available for carrying out creep tests in simple shear. Plazek (1968) has carried out extensive experiments on the creep testing of polymers over wide ranges of time and temperature.

*3.7.2 Dynamic methods: oscillatory strain*

The use of oscillatory methods increased considerably with the development of commercial rheogoniometers, and a further boost was given when equipment became available for processing the input and output signals to give in-phase and out-of-phase components directly. With modern instruments it is now possible to display automatically the components of the modulus as functions of frequency.

A general advantage of oscillatory tests is that a single instrument can cover a very wide frequency range. This is important if the material has a broad spectrum of relaxation times. Typically, the frequency range is from $10^{-3}$ to $10^3$ s$^{-1}$. Hence a time spectrum from about $10^3$ to $10^{-3}$ s can be covered. If it is desired to extend the limit to longer times, static tests of longer duration than 3 hr ($10^4$ s) would be

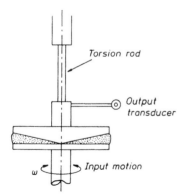

Fig. 3.13 Representation of the cone-and-plate apparatus for oscillatory tests. The specimen is positioned between the input motion and the output stress.

needed. The lower relaxation time limit of oscillatory methods can be extended by wave-propagation methods (see § 3.7.3).

The conventional oscillatory methods involve the application of either free or forced oscillatory strains in conventional tensile and shear geometries. An advantage possessed by the free vibration technique is that an oscillator is not required and the equipment can be fairly simple. On the other hand, the frequency range available is no more than two decades. The reason for this is that a change of frequency relies on a change in moment of inertia of the vibrating system and the scope for this is limited. The method is readily adaptable to torsional deformation with solid-like materials.

The wide frequency range quoted above is achieved with forced oscillations. We show in Fig. 3.13 the most common example of the forced-oscillation experiment, although the geometry could equally well be a parallel-plate or concentric-cylinder configuration. The test material is contained between a cone and plate, with the angle between the cone and plate being small ($<4°$). The bottom member undergoes forced harmonic oscillations about its axis and this motion is transmitted through the test material to the top member, the motion of which is constrained by a torsion bar. The relevant measurements are the amplitude ratio of the motions of the two members and the associated phase lag. From this information it is relatively simple to determine the dynamic viscosity $\eta'$ and the dynamic rigidity $G'$, measured as functions of the imposed frequency (see Walters 1975 for the details of this and related techniques).

### 3.7.3 Dynamic methods: wave propagation

A number of books are available which describe in detail the theory and practice of wave-propagation techniques. Kolsky (1963) has dealt with the testing of solids, Ferry (1980) has reviewed the situation as regards polymers and Harrison (1976) has covered liquids. The overall topic is usefully summarized by Whorlow (1980).

Basically, the waves are generated at a surface of the specimen which is in contact with the wave generator and the evaluation of the viscoelastic functions requires the measurement of the velocity and the attenuation through the specimen. One significant advantage of wave-propagation methods is that they can be adapted to high frequency studies: they have been commonly used in the kHz region and higher, even up to a few hundred GHz. This is invaluable when studying liquids which behave in a Newtonian manner in other types of rheometer. Such liquids include, as a general rule, those with a molecular weight below $10^3$. They include most of the non-polymeric liquids. Barlow and Lamb have made significant contributions in this area (see, for example, Barlow et al. 1967).

### 3.7.4 Dynamic methods: steady flow

In the oscillatory experiments discussed above, instrument members are made to oscillate and the flow is in every sense unsteady. A relatively new group of instruments for measuring viscoelastic behaviour is based on a different principle. The flow in these rheometers is steady in the sense that the velocity at a fixed point in the apparatus is unchanging. (Such a flow is described in fluid dynamics as being "steady in an Eulerian sense".) However, the rheometer geometry is constructed in such a way that individual fluid elements undergo an oscillatory shear (so that the flow is "unsteady in a Lagrangian sense"). A typical example of such an instrument is the Maxwell orthogonal rheometer which is shown in Fig. 3.14 (Maxwell and Chartoff 1965). It comprises two parallel circular plates separated by a distance $h$, mounted on parallel axes, separated by a small distance $d$. One spindle is rotated at *constant* angular velocity $\Omega$. The other is free to rotate and takes up a velocity close to that of the first spindle.

The components of the force on one of the plates are measured using suitable transducers. In the interpretation of the data it is assumed that the angular velocity of the second spindle is also $\Omega$. It can then be readily deduced that individual fluid elements are exposed to a sinusoidal shear and that the components of the force on each plate (in the plane of the plates) can be directly related to the dynamic viscosity and dynamic rigidity.

The Maxwell orthogonal rheometer and other examples of the steady-flow viscoelastic rheometers are discussed in detail by Walters (1975).

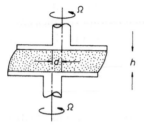

Fig. 3.14 Arrangement of rotating plates in a Maxwell orthogonal rheometer. Plate separation $h$; axis displacement $d$. One shaft rotates at constant velocity $\Omega$ and the second shaft takes up (nearly) the same velocity.

# CHAPTER 4

# NORMAL STRESSES

## 4.1 The nature and origin of normal stresses

We have already stated in §1.5 that, for a steady simple shear flow given by

$$v_x = \dot{\gamma} y, \quad v_y = v_z = 0, \tag{4.1}$$

the relevant stress distribution for non-Newtonian liquids can be expressed in the form

$$\left. \begin{aligned}
\sigma_{xy} &= \sigma = \dot{\gamma} \eta(\dot{\gamma}), \quad \sigma_{xz} = \sigma_{yz} = 0, \\
\sigma_{xx} - \sigma_{yy} &= N_1(\dot{\gamma}), \quad \sigma_{yy} - \sigma_{zz} = N_2(\dot{\gamma}).
\end{aligned} \right\} \tag{4.2}$$

The variables $\sigma$, $N_1$ and $N_2$ are sometimes called the viscometric functions. A useful discussion of the importance of these functions is given by Lodge (1974, p. 212). In this chapter, we shall be concerned with the normal stress differences $N_1$ and $N_2$ or, equivalently, the so-called normal stress coefficients $\Psi_1$ and $\Psi_2$, where

$$N_1 = \dot{\gamma}^2 \Psi_1, \quad N_2 = \dot{\gamma}^2 \Psi_2. \tag{4.3}$$

In principle, it is possible for a non-Newtonian *inelastic* model liquid to exhibit normal-stress effects in a steady simple shear flow. The so called Reiner–Rivlin fluid, which is a general mathematical model for an inelastic fluid (see §8.4), can be used to demonstrate this. However, all the available experimental evidence is that the theoretical normal stress distribution predicted by this model, viz. $N_1 = 0$, $N_2 \neq 0$ is not observed in any known non-Newtonian liquid. In practice, normal-stress behaviour is always that to be expected from models of viscoelasticity, whether they be mathematical or physical models.

The normal stress differences are associated with non-linear effects (cf. §1.3). Thus, they did not appear explicitly in the account of linear viscoelasticity in Chapter 3. In the experimental conditions of small-amplitude oscillatory shear, in which linear viscoelasticity is demonstrated and the parameters measured, the three normal stress components have the same value. They are equal to the ambient pressure, which is isotropic. Similarly, in steady flow conditions, provided the flow is slow enough for second-order terms in $\dot{\gamma}$ to be negligible, the normal stresses are again equal to the ambient pressure. As the shear rate is increased, the normal stress differences first appear as second-order effects, so that we can write

$$
\left. \begin{aligned}
N_1 &= A_2 \dot{\gamma}^2 + O(\dot{\gamma}^4), \\
N_2 &= B_2 \dot{\gamma}^2 + O(\dot{\gamma}^4),
\end{aligned} \right\} \tag{4.4}
$$

where $A_2$ and $B_2$ are constants and, as implied, the normal stress differences are even functions of the shear rate $\dot{\gamma}$. (The mathematically-minded reader may confirm this expectation by undertaking a simple analysis for the hierarchy equations given in Chapter 8 (eqns. 8.23–8.25), which are argued to be generally valid constitutive equations for sufficiently slow flow).

From a physical point of view, the generation of unequal normal stress components, and hence non-zero values of $N_1$ and $N_2$, arises from the fact that in a flow process the microstructure of the liquid becomes anisotropic. For instance, in a dilute polymeric system, the chain molecules, which at rest occupy an enveloping volume of approximately spherical shape, deform to an ellipsoidal shape in a flow field. The molecular envelope before and during deformation is shown in Fig. 4.1. The droplets in an emulsion change shape in a similar way. In the polymeric system at rest, entropic forces determine the spherical shape whilst the requirement of a minimum interfacial free-energy between an emulsion droplet and the surrounding liquid ensures practically spherical droplets in the emulsion at rest. It follows therefore that restoring forces are generated in these deformed microstructures and, since the structures are anisotropic, the forces are anisotropic. The spherical structural units deform into ellipsoids which have their major axes tilted towards the direction of flow. Thus the restoring force is greater in this direction than in the two orthogonal directions. The restoring forces give rise to the normal stress components of eqns. 4.2. It can be appreciated, from this viewpoint of their origin, why it is that the largest of the three normal stress components is always observed to be $\sigma_{xx}$, the component in the direction of flow. According to the principles of continuum mechanics, the components can have any value, but it would be an unusual microstructure that gave rise to components whose relative magnitudes did not conform to $N_1 \geqslant 0$, i.e. $\sigma_{xx} \geqslant \sigma_{yy}$. It is conceivable that a strongly-orientated rest-structure, as found in liquid crystals, might produce such unusual behaviour in certain circumstances.

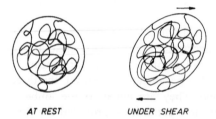

AT REST                    UNDER SHEAR

Fig. 4.1 The molecular envelope before and during shear deformation.

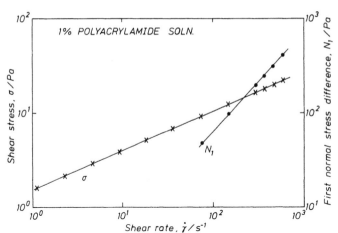

Fig. 4.2 Viscometric data for a 1% aqueous solution of polyacrylamide (E10 grade). 20 °C. Note that, over the shear-rate range $10^2$ to $10^3$ s$^{-1}$, $N_1$ is about ten times larger than $\sigma$.

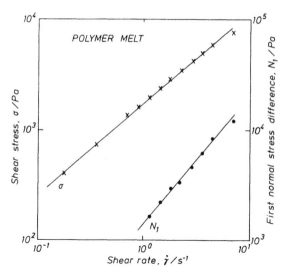

Fig. 4.3 Viscometric data for a polypropylene copolymer. 230 °C. Note that, over the shear-rate range $10^0$ to $10^1$ s$^{-1}$, $N_1$ is comparable in magnitude with $\sigma$.

## 4.2 Typical behaviour of $N_1$ and $N_2$

In view of the discussion in the previous section of the thermodynamic origin of the normal stresses, we expect $N_1$ to be a positive function of shear rate $\dot{\gamma}$. All reliable experimental data for elastic liquids are in conformity with this and show positive values of $N_1$ for all shear rates. Figures 4.2 and 4.3 show typical examples for a polymer solution and polymer melt, respectively. Note that $N_1$ may have a

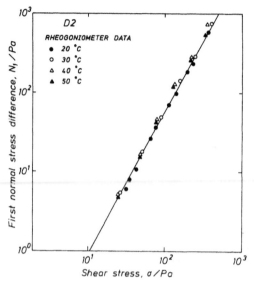

Fig. 4.4 A plot of ln $N_1$ against ln $\sigma$ at various temperatures for the polymer solution D2, which is a 10% w/v solution of polyisobutylene (Oppanol B50) in dekalin (cf. Lodge et al. 1987).

power-law behaviour over a range of shear rates, and we could write (cf. the related discussion concerning the shear stress $\sigma$ in §2.3.2)

$$N_1 = A\dot{\gamma}^m, \tag{4.5}$$

where $A$ and $m$ are constants, with $m$ being typically in the range $1 < m \leqslant 2$. As with the shear stress, the power-law region cannot extend to very low shear rates (unless $m = 2$).

It is clear from Fig. 4.2 that the normal stress difference $N_1$, in this case, is higher than the shear stress $\sigma$ and such an observation is not unusual. The ratio of $N_1$ to $\sigma$ is often taken as a measure of how elastic a liquid is; specifically $N_1/(2\sigma)$ is used and is called the *recoverable shear*. It follows that recoverable shears greater than 0.5 are not uncommon in polymeric systems. They indicate a 'highly elastic' state.

For polymeric systems, it is often found that a plot of ln $N_1$ against ln $\sigma$ for a range of temperatures results in a unique relationship which is a reasonably straight line of slope near 2 (see also §6.10 and Fig. 6.12). Figure 4.4 gives an illustration of such behaviour for a polymer solution. A straight line of slope 2 is expected at low shear rates in the so-called second-order region but there is no fundamental justification for the line to be independent of temperature. Nor is there any fundamental justification for this unique relationship to apply, as it often does, outside the second-order regime.

It is generally conceded that the second normal stress difference $N_2$ is small in

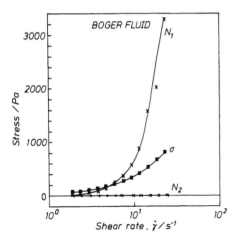

Fig. 4.5 Viscometric data for a Boger fluid: 0.184% polyisobutylene in a mixture of kerosene and polybutene (B.P. Hyvis 30). 25 °C.

comparison to $N_1$. Indeed, for a so-called Boger fluid *, the second normal stress difference has been found to be virtually zero (see, for example, Fig. 4.5 and Keentok et al. 1980). We also remark that in the early days of normal stress measurement (c.1950) $N_2 = 0$ was known as the Weissenberg hypothesis, and, within the limitations of the first-generation rheometers, experimental results on a number of systems were found to be in reasonable agreement with the hypothesis; the first such test being made by Roberts (1953) with a prototype version of the Weissenberg rheogoniometer. It is also noteworthy that some of the simpler microrheological models for polymeric systems predict $N_2 = 0$ (cf. Chapter 6). With $N_1 > 0$, $N_2 = 0$, we note that the resulting normal stress distribution is equivalent to an extra tension along the streamlines, with an isotropic state of stress in planes normal to the streamlines.

Modern rheometers are capable of determining $N_2$ with a reasonable degree of precision, although the level of tolerance is not as high as that associated with the determination of $\sigma$ and $N_1$. Non-zero values of $N_2$ can now be detected and measured in many systems, but the ratio of $|N_2|$ to $N_1$ is usually small ($\leqslant 0.1$). Present reliable data on polymeric systems all show $N_2$ to be zero or negative: Fig. 4.6 shows the viscometric functions for a 2% solution of polyisobutylene in decalin (the so-called D1 liquid). A comprehensive "round-robin" series of experiments was carried out on D1 and the findings are given by Walters (1983) and Alvarez et al. (1985). In this round-robin work, different types of instrument were used as well as different observers. The excellent agreement between the results shows that with

---

* A Boger (1977(a)) fluid is a very dilute solution ($\sim 0.1\%$) of a high molecular-weight polymer in a very viscous solvent. Although the solution does, in fact, display shear-thinning, the fall in viscosity is very small compared to the zero-shear value, and for practical purposes the viscosity appears to be constant (see also §7.2).

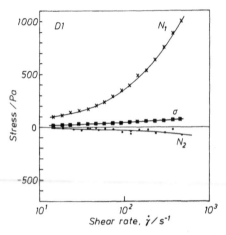

Fig. 4.6 Viscometric data for the polymer solution D1, which is a 2% w/v polyisobutylene (Oppanol B200) solution in dekalin. 25 °C (cf. Alvarez et al. 1985).

modern instruments it is possible to obtain consistency and accuracy in the measurement of normal stress components.

### 4.3 Observable consequences of $N_1$ and $N_2$

Normal stresses are responsible for a number of effects of laboratory interest and of commercial importance. Those included here by way of examples are observable with the aid of relatively simple equipment.

Perhaps the most well known and certainly the most dramatic effect is the rod-climbing phenomenon, usually referred to as the 'Weissenberg effect'. It is produced when a rotating rod is dipped into a squat vessel containing an elastic liquid. Whereas a Newtonian liquid would be forced towards the rim of the vessel by inertia, and would thus produce a free surface that is higher at the rim than near the rod, the elastic liquid produces a free surface that is much higher near the rod, as shown in Fig. 4.7. The observed rise of the surface is independent of the direction of rotation.

The Weissenberg effect may be viewed as a direct consequence of the normal stress $\sigma_{xx}$, which acts like a hoop stress around the rod. This stress causes the liquid to "strangle" the rod and hence move along it. The reaction of the bottom of the vessel, which should not be sited too far from the end of the rod, adds to the rise of the surface up the rod.

If the geometry of the rod-climbing experiment is changed by adding a flat disc to the end of the rod and aligning the disc to be close to, and parallel with, the bottom of the vessel, we have the configuration of one of a set of instruments used for measuring $N_1$ and $N_2$. Such instruments will be described later. Suffice it to say here that it can be shown that the strangulation caused by the first normal stress

Fig. 4.7 The Weissenberg effect shown by a solution of polyisobutylene (Oppanol B200) in polybutene (B.P. Hyvis 07). *Reproduced by permission of Shell Research Ltd.*

difference exerts a force between the bottom of the vessel and the disc, tending to push them apart. The measurement of this force can be used to yield normal stress information.

If the rod is replaced by a tube, open at both ends and with the disc (with a hole in the middle) still in place, the Weissenberg effect causes the elastic liquid to flow up the tube (Fig. 4.8). Flow will continue until the normal force is balanced by the gravitational force, provided there is enough liquid in the vessel. This is the principle of the "normal force pump", which is probably more of a novelty than a practical means of dispensing highly-elastic liquids.

Another phenomenon which can be reproduced with simple equipment, but yet has important consequences in manufacturing processes, is 'die swell', sometimes

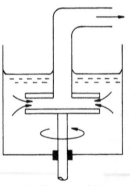

Fig. 4.8 The normal-force pump.

known as post-extrusion swelling. When an elastic liquid is extruded from a die or flows from the exit of a tube, it usually swells to a much greater diameter than that of the hole, as shown in Fig. 4.9. In fact, Newtonian liquids can also show die swell, but only at low rates of flow (with about a 13% swelling at negligibly small Reynolds numbers), and as the flow rate increases the swelling decreases, eventually becoming a contraction. In contrast, die swell of an elastic liquid increases as the

Fig. 4.9 Die swell shown by a solution of 1% polyacrylamide in a 50/50 mixture of glycerol and water.

flow rate increases. A swelling of up to two or three times the hole diameter is not unusual.

A convenient way of visualizing the origin of die swell is to consider the elastic liquid flowing towards the orifice as a bundle of elastic threads stretched by the $\sigma_{xx}$ normal stress component and when they emerge they are free to relax. The relaxation causes the threads to shorten in length, hence the bundle increases in diameter.

An important commercial process which is affected by die swell is the manufacture of rods, tubes and sheets of polymeric material. These articles are made by extrusion of molten polymer, which is an elastic liquid. Die swell causes problems in the control of the final thickness of the articles. The phenomenon is sensitive to the molecular-weight distribution of the polymer and such processing variables as flow rate and temperature. Increasing the length of the entry to the nozzle and reducing the angle of convergence to it are practical ways of reducing swelling, although at the expense of an increased pressure drop. However, die swell cannot be completely suppressed, so the satisfactory manufacture of a uniform product requires close control of the conditions.

Finally, we should mention that the measurement of the equilibrium amount of die swell produced under fully-controlled experimental conditions forms the basis of another method of measuring normal stress differences. There is a close link between this method and the method known as jet-thrust. In the latter, the force exerted by the emerging jet (of a necessarily mobile liquid), either directly onto an intercepting transducer or as a reaction on the flow tube, is related to the die swell and therefore also to normal stress levels (see, for example, Davies et al., 1975, 1977).

Normal stress effects are also important in those laminar mixing processes which involve disc impellers and may occur to some extent with other types. The flow pattern for a relatively inelastic liquid results from the interaction between viscous and inertial forces and comprises a radial outflow from the central impeller and return flows distant from the impeller as shown in Fig. 4.10. However, for a highly elastic liquid, the direction of flow can be completely reversed. There are intermediate cases when both types of flow pattern coexist. In this situation, the flow pattern characteristic of the elastic liquid and generated by the normal stresses hugs the impeller whilst the inelastic-liquid pattern is found in regions remote from the impeller. Examples are given in Fig. 4.10. Obviously, liquid contained in the one pattern will not mix very well with liquid in the other. Which type of flow pattern is obtained in a given situation depends on the ratio of the elastic forces to the inertial forces. This ratio is a dimensionless group which is given by the ratio of the Weissenberg number to the Reynolds number $(W_e/R_e)$ *. The dual flow pattern is to be found at intermediate values of this ratio.

---

* The 'Weissenberg number' $W_e$ may be defined as *the ratio of the first normal stress difference to the shear stress in a steady simple shear flow.*

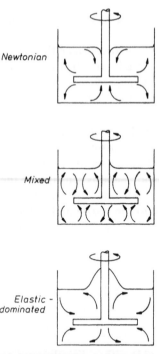

Fig. 4.10 Normal-stress effects in mixing; showing how the form and direction of the circulation are affected by increasing elasticity.

Although the second normal stress difference $N_2$ is generally of far less practical significance than the first normal stress difference $N_1$, it is important to point out that in some situations $N_2$ is very important. For example, it is the function $N_2(\dot{\gamma})$ which determines whether or not rectilinear flow in a pipe of non-circular cross section is possible (see, for example, Townsend et al. 1976). A related problem is wire coating, and the importance of $N_2$ in this practical problem has been stressed by Tadmor and Bird (1974).

Such examples of the importance of $N_2$ are rare and apart from those who are directly concerned with the situations cited, most practitioners in non-Newtonian fluid mechanics tend to confine attention to $N_1$, especially in view of the relative difficulty of measuring $N_2$.

### 4.4 Methods of measuring $N_1$ and $N_2$

As a generality, the ideal method of measuring normal stress differences would involve an uncomplicated shear geometry, which could be easily made, be amenable to an exact mathematical analysis, and would enable the normal stress differences to be measured separately from shear and inertial forces. It is not possible to achieve the ideal, but the methods outlined below are the nearest to it.

We have already stated that $N_1$ is large in comparison to $N_2$ and that the latter is the more difficult to measure. For these reasons, it is customary to give greater emphasis to methods for $N_1$ determination, and, in routine laboratory work, to confine measurements to $N_1$.

The simplest flow is that shown in Fig. 1.1. in connection with Newton's postulate. It can be generated by sliding two parallel plates over each other ('plane Couette flow'). There has been a limited number of attempts to use this method, but it has practical limitations. Mobile liquids may run out of the gap, and it is impossible to maintain a continuous shear for very long; hence the method is restricted to extremely viscous liquids or (for less-viscous liquids) to very narrow gaps, hence high shear rates. In this form of the parallel-plate apparatus, Dealy (see, for example Dealy and Giacomin 1988) uses flow birefringence as the means of measuring normal stress and he inserts a flush-mounted transducer into the surface of one plate to measure the shear stress, free from edge effects.

In view of the limitations of apparatus constructed for generating the primitive simple shear, it is not surprising that a detailed search has taken place for flows which are equivalent to steady simple-shear flow in a well-defined mathematical sense. These 'viscometric flows' include 'Poiseuille flow' (i.e. steady flow under a constant pressure gradient in a pipe of circular cross section), 'circular Couette flow' (i.e. steady flow between coaxial cylinders in relative rotation), torsional flow (i.e. steady flow between parallel plates, one of which rotates about a normal axis) and the corresponding cone-and-plate flow, which will figure prominently in the following discussion. This list is not meant to be exhaustive and the reader is referred to Walters (1975), Dealy (1982) and Lodge (1974) for other examples and greater detail.

The proof that all these flows are equivalent to steady simple-shear flow (with the stress distribution expressible in terms of $\sigma$, $N_1$ and $N_2$) is non-trivial and has been approached from different standpoints by Lodge (1974), Coleman et al. (1966), Bird et al. (1987(a) and (b)) and Walters (1975).

### 4.4.1 Cone-and-plate flow

It is probably true to say that the cone-and-plate geometry is the most popular for determining the normal stress differences. The basic geometry is shown schematically in Fig. 4.11 (see also Chapter 2). The test liquid is contained between a

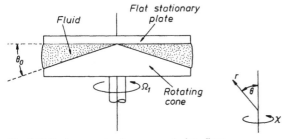

Fig. 4.11 Basic geometry for cone-and-plate flow.

rotating cone and a flat stationary plate. (Alternatively the plate is designed to rotate with the cone stationary, with a small advantage as regards alignment).

With respect to suitably chosen spherical polar coordinates, the physical components of the velocity vector at any point in the liquid are assumed to be (see Fig. 4.11)

$$v_{(r)} = 0, \quad v_{(\theta)} = 0, \quad v_{(x)} = r \sin\theta \, \Omega(\theta), \tag{4.6}$$

with the boundary conditions that the angular velocity is zero at the plate surface and $\Omega_1$ at the surface of the cone, i.e.

$$\Omega(\pi/2) = 0, \quad \Omega[(\pi/2) + \theta_0] = \Omega_1, \tag{4.7}$$

where $\theta_0$ is the gap angle. It can be shown that the flow represented by eqn (4.6) is equivalent to a steady simple-shear flow with shear rate $\dot{\gamma} = \sin\theta \, d\Omega/d\theta$, and that, when the stress equations of motion are taken into account, we obtain for the shear stress $\sigma$ (see, for example, Walters 1975, Chapter 4):

$$\sigma(\dot{\gamma}) = \bar{A} \, \mathrm{cosec}^2\theta, \tag{4.8}$$

and, for the normal stress differences,

$$2\rho r^2 \sin^2\theta \, \Omega \frac{d\Omega}{d\theta} = \frac{d}{d\theta} [N_1(\dot{\gamma}) + 2N_2(\dot{\gamma})], \tag{4.9}$$

where $\rho$ is the density and $\bar{A}$ is a constant to be determined from the boundary conditions. Equations (4.8) and (4.9) are in general incompatible (in the sense that a solution to (4.8) will not be a solution to (4.9) and vice versa), unless we make the following assumptions:

(*i*) inertial effects are negligible, which means setting $\rho = 0$ in (4.9);

(*ii*) the angle between the cone and the plate is small enough to allow us to set $\mathrm{cosec}^2\theta = 1$ in (4.8), which in practical terms means restricting the gap angle $\theta_0$ to be no greater than 4°.

With assumptions (*i*) and (*ii*), we have

$$\dot{\gamma} = \Omega_1/\theta_0, \tag{4.10}$$

i.e. there is a constant shear rate throughout the sample and it is independent of the form of the viscometric functions. Equations (4.8) and (4.9) are now compatible.

It is easy to show that the torque $C$ acting on the stationary plate of radius $a$ is given by (cf. Chapter 2)

$$C = \frac{2\pi a^3}{3} \sigma(\dot{\gamma}), \tag{4.11}$$

and that if $\bar{p}$ is the pressure on the plate at a radius $r$, in excess of that due to atmosphere pressure, then

$$\frac{d\bar{p}}{d(\ln r)} = -[N_1(\dot{\gamma}) + 2N_2(\dot{\gamma})], \tag{4.8}$$

i.e. there is a logarithmic dependence of $\bar{p}$ on $r$ and the slope of the $(\bar{p}, \ln r)$ curve can be used to yield $N_1(\dot{\gamma}) + 2N_2(\dot{\gamma})$. Further, if the pressure is integrated over the plate, we obtain the total normal force $F$ on the plate and it can then be shown that (Walters 1975, Chapter 4)

$$F = \frac{\pi a^2}{2} N_1(\dot{\gamma}). \tag{4.9}$$

This force acts in the direction of the axis of rotation and pushes the cone and plate apart. It is essentially the same force that produces the Weissenberg rod-climbing effect.

The above analysis tells us that the measurement of the rotational speed will give the shear rate and that measurement of the torque on the stationary plate will give the shear stress. As regards the normal stress differences, there are two alternatives. First, the force $F$ gives $N_1$; secondly the radial distribution of pressure gives $N_1 + 2N_2$. Hence, in principle, the two normal stress differences can be obtained if these two alternatives are both used.

There is a basic conflict in the normal force measurement, since the force $F$ tends to separate the cone from the plate. The consequence of such a separation, if it were allowed, is to upset the condition of uniform shear rate throughout the sample and to reduce the mean value of shear rate. The ideal measuring system should be rigid to axial forces. For systems which are not rigid, a servo-mechanism is used to maintain the cone-plate gap.

Various potential sources of error have to be borne in mind when performing experiments in the cone-and-plate geometry. The more important are enumerated below and we refer the reader to the texts of Walters (1975) and Whorlow (1980) for further details.

### I Inertial effect

The origin and nature of the effect of inertia has already been mentioned. It gives rise to the so-called "negative normal stress effect", whereby the plates are pulled together and the measured value of the force $F$ is smaller than the true value. The reduction in the force $F$ is given by (Walters 1975)

$$\Delta F = 3\pi \rho \Omega_1^2 a^4 / 40. \tag{4.10}$$

This formula is used to correct experimental values: it can be seen that it is sensitive to the rotational speed and very sensitive to the plate radius.

## II Hole-pressure error

A major source of error which can arise when the pressure-distribution method is used is known as the hole-pressure error (Broadbent et al. 1968). Any method of measuring pressure which relies on the use of a hole in the bounding surface gives a low result with elastic liquids owing to the stretching of the flow lines as they pass over the hole. The reduction is directly related to $N_1$ and is in fact used as a means of measuring $N_1$. This method is described later in this chapter, where a more detailed description is given. The error is avoided by the use of stiff, flush-mounted pressure transducers.

## III Edge effects

"Shear fracture" places an upper limit on the usable shear rate range for highly elastic materials like polymer melts. It is observed as a sharp drop in all stress components, and at the same time a change in shape of the free surface can be seen, as well as a rolling motion in the excess liquid around the rim of the plates. A horizontal free surface forms in the test sample at the rim and grows towards the centre, hence reducing the sheared area. The limiting shear rate can be quite low, depending on the liquid and the cone dimensions. Expressed as a critical normal stress $N_1^{(c)}$, the limit is given by

$$N_1^{(c)} = c/a\theta_0,\tag{4.11}$$

where $c$ is a constant of the liquid. For a given liquid, shear fracture is minimized if the cone radius and gap angle are small.

The name "shear fracture" was given to the effect by Hutton (1965) owing to its similarity to 'melt fracture', which limits the occurrence of steady flow of polymer melts in tubes. Tordella (1956) made the first systematic study of melt fracture and noted that when the effect is severe the stream of melt breaks up with an accompanying tearing noise.

Another edge effect, also pointed out by Hutton (1972) is ascribable to changes in contact angle and/or surface tension of the test liquid brought about by shear. The effect is of potential importance when the test liquid possesses only small normal stresses.

## IV Miscellaneous precautions

The alignment of the cone axis to be coincident with the rotational axis, the setting of the cone tip in the surface of the plate, and the minimizing of, or correction for, viscous heating are other important matters to be taken into account in accurate work.

### 4.4.2 Torsional flow

Torsional flow is shown schematically in Fig. 4.12 (see also Chapter 2). Clearly, commercial instruments which are designed to work in a cone-and-plate mode can be easily adapted to the parallel-plate geometry and vice versa.

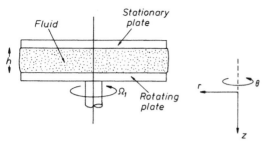

Fig. 4.12 Basic geometry for torsional flow.

In this case, with respect to suitably defined cylindrical polar coordinates, the velocity distribution can be taken to be

$$v_{(r)} = 0, \quad v_{(\theta)} = r\omega(z), \quad v_{(z)} = 0, \tag{4.12}$$

subject to the boundary conditions at the two plates

$$\omega(0) = 0, \quad \omega(h) = \Omega_1, \tag{4.13}$$

where $h$ is the gap between the plates.

Taking into account the fact that eqn. (4.12) is equivalent to a steady simple-shear flow, the stress equations of motion are satisfied, with the shear rate given by

$$\dot{\gamma} = r\Omega_1/h, \tag{4.14}$$

provided

$$2\rho r \frac{d\omega}{dz} = 0. \tag{4.15}$$

Equation (4.14) implies that the shear rate is independent of the viscometric functions: it depends on radial distance $r$, but is constant across the gap for fixed $r$. This time, we see from eqn. (4.15) that we have to neglect inertia for compatibility and there is no essential restriction on the gap $h$, except of course that this must not be too large that edge effects in a practical rheometer become important. The edge effects mentioned in connection with the cone-and-plate instrument apply here.

After some routine mathematics, it is possible to show that the viscosity function can be determined from measurements of the torque $C$ through the equation (cf. Chapter 2):

$$\eta(\dot{\gamma}_a) = \frac{3C}{2\pi a^3 \dot{\gamma}_a} \left( 1 + \frac{1}{3} \frac{d \ln C}{d \ln \dot{\gamma}_a} \right), \tag{4.16}$$

where $\dot{\gamma}_a$ is the shear rate at the rim ($r = a$). It can also be shown that

$$(N_1 - N_2)|_{\dot{\gamma}_a} = \frac{2F}{\pi a^2} \left( 1 + \frac{1}{2} \frac{d \ln F}{d \ln \dot{\gamma}_a} \right), \tag{4.17}$$

where $F$ is again the total normal force on the plates. We see that total-force data yield the combination $N_1 - N_2$ at the shear rate $\dot{\gamma}_a$ at the rim. Clearly, total force measurements taken in the cone-and-plate and parallel-plate geometries can be combined to yield $N_1$ and $N_2$ separately. However, since $N_2$ is small and two separate experiments have to be performed, significant scatter in the final data can be anticipated unless there is very refined experimentation.

Of interest is the fact that, in principle, relatively high shear rates can be attained with small gaps $h$. This has been utilized in the so-called "torsional balance rheometer" of Binding and Walters (1976) to obtain normal stress data at shear rates in excess of $10^4$ s$^{-1}$. In this form of the instrument, a predetermined external normal force is applied to the upper plate and the separation $h$ is allowed to vary until this force balances the normal force generated by the liquid. Gap $h$ is measured and eqns. (4.16) and (4.17) applied.

### 4.4.3 Flow through capillaries and slits

A consistent theory for normal stress measurement in flow through a capillary is available (cf. Walters 1975, Chapter 5), but this depends critically on the flow being "fully developed" at the exit to the capillary, by which we mean that the Poiseuille flow generated away from the influence of end effects should be maintained right up to the capillary exit with no rearrangement of the velocity profile. Furthermore, the experimental results have to be carried out with flush-mounted pressure tranducers to avoid the hole-pressure error problem, and this is difficult on the curved walls of a capillary. Therefore, the use of the potentially attractive exit-pressure measurement technique of determining normal stress data is controversial (cf. Boger and Denn 1980). The associated jet-thrust technique for low viscosity elastic liquids and high shear rates is also based on the assumption of fully developed flow at the capillary exit (cf. Davies et al. 1975, 1977).

In the Lodge (1988) stressmeter, which uses pressure-driven flow through a slit, the hole-pressure error mentioned above is turned to good use in the measurement of $N_1$. As liquid flows past the hole under an applied pressure gradient, streamlines adjacent to the boundary wall are deviated into and then out of the hole, as shown in Fig. 4.13. For an elastic liquid the deviation is viewed as a stretching by the normal stress component acting along the streamlines. The net result is a lowering of the pressure in the hole. The holes in the stressmeter are a pair of slots set across the flow direction as shown in Fig. 4.13. The reduction in pressure $\Delta p$ is measured

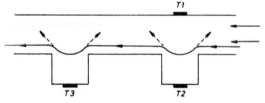

Fig. 4.13 Schematic diagram of the Lodge stressmeter for normal stress and shear stress measurement. The dotted lines represent the tension in the streamlines resulting in a lowered pressure in the holes.

(as $p_1 - p_2$) between the bottom of the hole beneath one of the slots and a flush-mounted transducer whose diaphragm forms the boundary wall on the opposite side of the slit. The difference in pressure in the two slots ($p_2 - p_3$) gives the shear stress.

It has been shown theoretically that $\Delta p$ is given by (cf. Tanner and Pipkin 1969)

$$\Delta p = N_1/4 \qquad\qquad\qquad\qquad (4.18)$$

for a second-order simple-fluid model (which will be shown in §8.5 to be a valid slow-flow approximation for a general class of elastic liquid). The generalization of eqn. (4.18) embodied in the so called HPBL equation is used to interpret results for flows which are certainly outside the "slow-flow" regime. The interpretation of results is accordingly based on what must be seen as an empirical equation, with no theoretical justification except at low shear rates. However, it appears to work well, judged by recent comparative studies with other instruments (Lodge et al. 1987), and shear rates as high as $10^6$ s$^{-1}$ have been reached with multigrade motor oils with this technique.

### 4.4.4 Other flows

Tanner (1970) has proposed that the free surface shape in gravity-driven flow down an open tilted trough can be used to calculate the second normal stress difference $N_2$. In general terms, if the free surface rises near the centre, $N_2$ is negative and, if it falls, $N_2$ is positive. The interpretation of data is not trivial, but the technique provides a convenient method of determining estimates of $N_2$ at low shear rates.

Circular Couette flow between rotating cylinders is popular in the determination of the viscosity of non-Newtonian liquids (cf. §2.4). Attempts have also been made to employ the flow to determine normal stress information from pressure readings (Broadbent and Lodge 1972). The fact that the technique has not been popular with experimentalists since the original paper is probably an indication of the difficulty of using the technique, or it at least points to the fact that much easier methods are available in rotational rheometry using, for example, cone-and-plate flow.

## 4.5 Relationships between viscometric functions and linear viscoelastic functions

Earlier in this chapter we argued that normal stress differences in a simple-shear flow were a direct consequence of viscoelasticity. We recall from §3.5 that viscoelasticity can also be studied through a small-amplitude oscillatory-shear flow, the resulting stress distribution for an elastic liquid being expressible in terms of the dynamic viscosity $\eta'$ and the dynamic rigidity $G'$. Now, since the departure from a Newtonian response in the viscometric functions $\eta$, $N_1$ and $N_2$ and in the dynamic functions $\eta'$ and $G'$ can be ascribed to viscoelasticity, we should not be surprised to find that there are relationships between the various rheometrical functions. In fact,

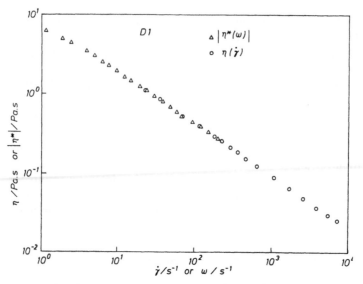

Fig. 4.14 The Cox–Merz rule applied to the polymer solution D1, which is a 2% w/v polyisobutylene (Oppanol B200) solution in dekalin. 25 °C.

it is not difficult to deduce the exact relationships in the lower limits of frequency and shear rate:

$$\eta'(\omega)\big|_{\omega \to 0} = \eta(\dot{\gamma})\big|_{\dot{\gamma} \to 0}, \tag{4.19}$$

$$\frac{G'(\omega)}{\omega^2}\bigg|_{\omega \to 0} = \frac{N_1(\dot{\gamma})}{2\dot{\gamma}^2}\bigg|_{\dot{\gamma} \to 0} = \frac{\Psi_1(\dot{\gamma})}{2}\bigg|_{\dot{\gamma} \to 0}. \tag{4.20}$$

The former relationship states that the viscosity measured in oscillatory shear in the zero-frequency limit is equal to the low shear viscosity measured in steady shear. Equation (4.20) is a relationship between the limiting values of dynamic rigidity and first normal stress difference.

In many cases, it is easier to carry out dynamic measurements than steady shear measurements and (4.19) and (4.20) provide a means of estimating the levels of $\eta$ and $\Psi_1$ (and hence $N_1$) from measurements of $\eta'$ and $G'$.

We note that in view of eqn. (4.19) and the fact that both $\eta$ and $\eta'$ are usually monotonic decreasing functions of $\dot{\gamma}$ and $\omega$, respectively, various attempts have been made to develop empirical relationships between $\eta$ and $\eta'$ at other than the lower limits of shear rate and frequency. The most popular, and most successful in this respect, certainly for polymeric liquids, is the so-called Cox–Merz (1958) rule, which proposes that $\eta$ should be the same function of $\dot{\gamma}$ as $|\eta^*|$ is of $\omega$, where $|\eta^*|$ is the modulus of the complex viscosity, i.e.

$$|\eta^*| = \left[(\eta')^2 + (G'/\omega)^2\right]^{1/2}. \tag{4.21}$$

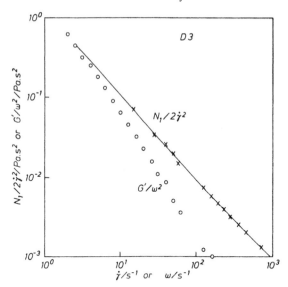

Fig. 4.15 A test of the relationship of eqn. (4.20) showing the asymptotic approach of the oscillatory and steady shear parameters. Steady shear and dynamic data for the polymer solution D3, which is a 1.5% w/v polyisobutylene (Oppanol B200) solution in dekalin. 20 °C.


In Fig. 4.14 we provide an example of the application of the Cox–Merz rule to a polymer solution.

In view of (4.20) and the fact that both $G'$ and $N_1$ are monotonic increasing functions of $\omega$ and $\dot{\gamma}$, respectively, we might be led to expect that a relationship analogous to the Cox–Merz rule will hold between $G'$ and $N_1$ (see, for example, §6.10 and cf. Al-Hadithi et al. 1988). The limiting relationship (4.20) has been confirmed many times and Fig. 4.15 provides just one example of this for a polymeric liquid, where we see that the values of $N_1/2\dot{\gamma}^2$ and $G'/\omega^2$ coincide at low values of $\dot{\gamma}$ and $\omega$.

CHAPTER 5

# EXTENSIONAL VISCOSITY

## 5.1 Introduction

The subject of 'extensional' (or 'elongational') flow received scant attention until the mid 1960s. Up to that time rheology was dominated by shear flows. In the last twenty years or so the situation has changed dramatically with the dual realization that extensional flow is of significant relevance in many practical situations and that non-Newtonian elastic liquids often exhibit dramatically different extensional flow characteristics from Newtonian liquids. Accordingly, interest in the subject has mushroomed and much effort is now expended in trying to measure the extensional viscosity of non-Newtonian liquids, whether they be "stiff" systems like polymer melts or "mobile" systems like dilute polymer solutions, suspensions and emulsions. The general subject is covered in the book by Petrie (1979) entitled "*Elongational Flows*". Petrie's book requires more than a passing acquaintance with mathematics to be fully appreciated, but there is sufficient general detail in the book to make it important reading for anyone requiring a thorough knowledge of the subject. The works of Dealy (1982), Cogswell (1981), Meissner (1983, 1985), Münstedt and Laun (1981, 1986) are also important sources of information for those whose direct concern is polymer melts.

Unlike the situation in steady simple shear and oscillatory shear (see Chapters 2–4) where the subjects are mature, the study of extensional flow is still evolving. This is reflected in the slightly different style of the present chapter.

For the velocity field (see Fig. 5.1(a))

$$v_x = \dot{\epsilon}x, \quad v_y = -\dot{\epsilon}y/2, \quad v_z = -\dot{\epsilon}z/2, \tag{5.1}$$

where $\dot{\epsilon}$ is a constant extensional strain rate, the corresponding stress distribution can be conveniently written in the form

$$\left. \begin{array}{l} \sigma_{xx} - \sigma_{yy} = \sigma_{xx} - \sigma_{zz} = \dot{\epsilon}\eta_E(\dot{\epsilon}), \\ \sigma_{xy} = \sigma_{xz} = \sigma_{yz} = 0. \end{array} \right\} \tag{5.2}$$

where $\eta_E$ is the (uniaxial) extensional viscosity. In general, it is a function of the extensional strain rate $\dot{\epsilon}$, just as the shear viscosity is a function of shear rate $\dot{\gamma}$ (§2.3). However, we shall see that the behaviour of the extensional viscosity function

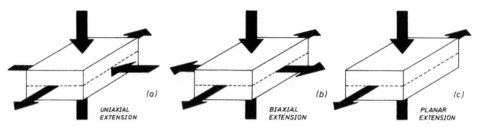

Fig. 5.1 The three different types of extensional flow fields are shown by the arrows: (a) Uniaxial; (b) Biaxial; (c) Planar.

is frequently qualitatively different from that of the shear viscosity. So, for example, highly elastic polymer solutions that possess a viscosity that *decreases* monotonically in shear (showing shear-thinning) often exhibit an extensional viscosity that *increases* dramatically with strain rate.

A fluid for which $\eta_E$ increases with increasing $\dot{\epsilon}$ is said to be *'tension-thickening'*, whilst, if $\eta_E$ decreases with increasing $\dot{\epsilon}$, it is said to be *'tension-thinning'*.

Experimentally, it is often not possible to reach the steady state implied in (5.1) and (5.2). Under these circumstances, it is convenient to define a transient extensional viscosity $\bar{\eta}_E(t, \dot{\epsilon})$, which is clearly a function of $t$ as well as $\dot{\epsilon}$. This arises from the obvious analogue to (5.2) given that the extensional flow field (5.1) is initiated at time $t = 0$ and maintained thereafter. In some respects this is a disappointing admission of difficulties which certainly do not normally occur in the measurement of the shear viscosity $\eta(\dot{\gamma})$. However, a study of $\bar{\eta}_E(t, \dot{\epsilon})$ can still throw considerable light on the rheological response of non-Newtonian liquids. It is also not without its industrial relevance, since in many practical situations liquids are exposed to extensional flow fields over a limited period of time only (see, for example, Bird et al. 1987(a) and Laun and Schuch 1988).

Another type of extensional deformation is the so-called *biaxial* extension, given by (see Fig. 5.1(b))

$$v_x = \dot{\epsilon}x, \quad v_y = \dot{\epsilon}y, \quad v_z = -2\dot{\epsilon}z, \tag{5.3}$$

where $\dot{\epsilon}$ is a constant. This type of extension is equivalent to stretching a thin sheet of material in two orthogonal directions simultaneously, with a corresponding decrease in the sheet thickness. It is found (approximately) when a circular free jet impinges on a flat plate or in a lubricated squeeze-film flow (see, for example, Soskey and Winter 1985) and when a balloon is inflated. The stress field corresponding to (5.3) can be written in the form

$$\left. \begin{array}{l} \sigma_{zz} - \sigma_{xx} = \sigma_{zz} - \sigma_{yy} = -\dot{\epsilon}\eta_{EB}(\dot{\epsilon}), \\ \sigma_{xy} = \sigma_{xz} = \sigma_{yz} = 0, \end{array} \right\} \tag{5.4}$$

where $\eta_{EB}$ is the *biaxial* extensional viscosity. It can be shown that (Walters 1975, p. 211)

$$\eta_{EB}(\dot{\epsilon}) = 2\eta_E(-2\dot{\epsilon}).\tag{5.5}$$

Finally, a two-dimensional *planar* extensional flow given by (see Fig. 5.1(c))

$$v_x = \dot{\epsilon}x, \quad v_y = -\dot{\epsilon}y, \quad v_z = 0,\tag{5.6}$$

where $\dot{\epsilon}$ is a constant, yields a *planar* extensional viscosity $\eta_{EP}$:

$$\sigma_{xx} - \sigma_{yy} = \dot{\epsilon}\eta_{EP}(\dot{\epsilon})\tag{5.7}$$

This type of extension is equivalent to stretching a thin flat sheet of material in one direction only (the $x$ direction), with a corresponding contraction in its thickness in the $y$ direction, but with no change in the width of the sheet. Planar extensional flow can be shown to be equivalent to that generally known as "pure shear" (see, for example, Walters 1975, Chapter 7).

In the present book, our general concern will be the *uniaxial* extensional viscosity $\eta_E$ and its comparison to the equivalent shear viscosity. Fuller details about subjects not enlarged on here are provided in the texts of Petrie (1979), Walters (1975), Tanner (1985) and Bird et al. (1987(a) and (b)).

## 5.2 Importance of extensional flow

In polymer processing (see §6.11.1), a case can be made out that some operations involve a significant component of extensional flow, with the obvious conclusion that the measurement of extensional viscosity may sometimes be as important as the determination of the shear viscosity. This will become increasingly so as manufacturers attempt to further increase production rates.

Any reasonably abrupt change in geometry in a processing operation will generate a flow with an extensional component and, in particular, flows through a sudden contraction or out of an orifice often lead to flow characteristics which cannot be predicted on the basis of shear viscosity alone. The polymer engineer must, therefore, have a working knowledge of extensional flow and must, if possible, know whether the materials he is processing are tension-thinning or tension-thickening. Certainly, the 'spinnability' of a polymeric liquid can be very dependent on its extensional viscosity behaviour. To illustrate this, consider the fibre-spinning process shown schematically in Fig. 5.2. It is clearly important for the process, which is dominated by extensional flow, to be stable and for the threadline not to snap. The tension along the threadline is obviously chosen to prevent fracture under normal operating conditions and the main concern is with the propagation and magnification of small disturbances, which are to some extent unavoidable in a physical process of this sort.

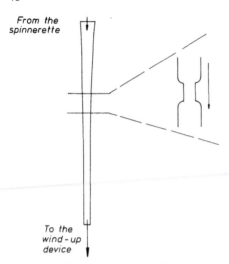

Fig. 5.2 The filament-necking imperfection in fibre spinning.

The stability of a spinning threadline is a vast area of study (see, for example, Petrie 1979) and we shall do no more than isolate one possible cause for concern. Let us speculate that for some reason a change in diameter occurs. From simple continuity considerations we would expect the narrower part of the filament to move faster than the rest of the threadline. Put in another way, the extensional strain rate will now be higher in the narrow part. If the polymeric liquid is tension-thinning, the resistance to extension is reduced in the narrow part and motion in this part of the threadline is further accelerated. It becomes thinner and may ultimately break.

If, on the other hand, the polymeric liquid is tension-thickening, the resistance in the narrow part of the filament will now be increased. The flow in the filament will slow down, the radius will increase and may be expected to return to that of the remainder of the threadline. Tension-thickening is therefore a stabilizing influence in this process.

Other examples of the importance of extensional viscosity in polymer processing could be cited. Certainly, the polymer engineer needs to be aware of the fact that two polymeric liquids which may have essentially the same behaviour in shear can show a different response in extension.

Later in this chapter we shall see that there is sufficient theoretical and experimental evidence available to support the view that very dilute solutions of flexible polymers can have extremely high extensional viscosities. Certainly, these can be orders of magnitude higher than those expected on the basis of Newtonian theory. This has important consequences in a number of practically important situations. For example, it may significantly affect the pressure losses encountered in polymer flooding in "enhanced oil recovery" (EOR) (see §6.11.3). Further, it may be the cause of the phenomenon known as "drag reduction". When small concentrations

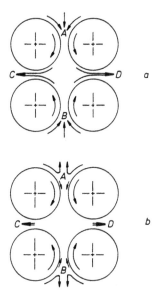

Fig. 5.3 The 4-roll mill experiment. Schematic representation of the fluid velocities for: (a) A Newtonian liquid; (b) A highly elastic liquid, for which high extensional stresses at A and B reduce the inflow.

(of the order of a few parts-per-million) are added to a Newtonian solvent like water, there is often a substantial reduction in drag in turbulent flow (Tanner 1985, p. 423). Drag reduction is of potential importance in many spheres. For example, small quantities of polymer may be injected into sewers during heavy rain to upgrade flow and so prevent flooding.

Many different mechanisms have been proposed to account for the phenomenon of drag reduction, but it may be linked to extensional viscosity and in particular to the suppression of the roll-wave motion and vortex stretching in the sublayer by the high extensional viscosity.

The potential importance of extensional viscosity effects in such processes as calendering and paper coating is suggested by the four-roll mill experiments discussed by Metzner and Metzner (1970) (see Fig. 5.3).

When the four-roll mill is immersed in a Newtonian liquid, the expected flow regime shown in Fig. 5.3(a) is observed. However, in the case of some dilute polymer solutions, the flow shown in Fig. 5.3(b) more adequately reflects the observations. The flow at A and B is too "strong" to permit substantial amounts of fluid to enter as in Fig. 5.3(a). Specifically, the anticipated flow has a high extensional component which results in high extensional stresses. The shear stresses generated by the rotating rollers are not strong enough to overcome the large extensional stresses and a reversed flow results at A and B.

Further examples of anomolous flow characteristics caused by high extensional viscosities are given in §5.6.

Finally, we remark that extensional flow experiments can be viewed as providing critical tests of any proposed constitutive equations. Indeed, the historical convention of matching only *shear* flow data with theoretical predictions in constitutive modelling may have to be rethought in those areas of interest where there is a large extensional contribution. It may be more profitable to match any *extensional* viscosity data which may be available, even if this means that the resulting constitutive model loses some of its predictive value so far as shear data are concerned.

### 5.3 Theoretical considerations

Continuum mechanics is able to provide some useful insights into the extensional-viscosity behaviour of non-Newtonian liquids. For example, the following limiting relations between extensional and shear viscosities are true (cf. Walters 1975, Petrie 1979)

$$\eta_E(\dot{\epsilon})\big|_{\dot{\epsilon}\to 0} = 3\eta(\dot{\gamma})\big|_{\dot{\gamma}\to 0},\tag{5.8}$$

$$\eta_{EP}(\dot{\epsilon})\big|_{\dot{\epsilon}\to 0} = 4\eta(\dot{\gamma})\big|_{\dot{\gamma}\to 0}.\tag{5.9}$$

We note that these relationships are valid for all values of $\dot{\epsilon}$ and $\dot{\gamma}$ in the case of Newtonian liquids. In particular, for Newtonian liquids,

$$\eta_E = 3\eta,\tag{5.10}$$

a result obtained by Trouton as early as 1906. Accordingly, rheologists have introduced the concept of the 'Trouton ratio' $T_R$ defined as

$$T_R = \frac{\eta_E(\dot{\epsilon})}{\eta(\dot{\gamma})}.\tag{5.11}$$

Elastic liquids are noted for having high Trouton ratios, but the definition as given in eqn. (5.11) is somewhat ambiguous, since it depends on both $\dot{\epsilon}$ and $\dot{\gamma}$, and some convention has therefore to be adopted to relate the strain rates in extension and shear. To remove this ambiguity and at the same time provide a convenient estimate of viscoelastic effects, Jones et al. (1987) have proposed the following definition, based on a simple analysis for an inelastic non-Newtonian fluid:

$$T_R(\dot{\epsilon}) = \frac{\eta_E(\dot{\epsilon})}{\eta(\sqrt{3}\,\dot{\epsilon})},\tag{5.12}$$

i.e., in the denominator, the shear viscosity is evaluated at the shear rate numerically equal to $\sqrt{3}\,\dot{\epsilon}$. Jones et al. show that if the flow is inelastic, $T_R$ is 3 *for all values of* $\dot{\epsilon}$.

They argue that any departure from the value 3 can be associated unambiguously with viscoelasticity, so that the definition (5.12) not only removes the ambiguity in the definition of the Trouton ratio, but also provides a convenient means of estimating viscoelastic response.

Note that variable (shear) viscosity effects are accommodated in the analysis of Jones et al., which illustrates convincingly that a fluid that is shear-thinning must also be expected to be tension-thinning in extension, *if viscoelastic effects are negligible or very small.*

Continuum mechanics also supplies a limiting relationship between the extensional viscosity $\eta_E$ and the normal stress coefficients $\Psi_1$ and $\Psi_2$ as determined in shear flow. A simple analysis for the so-called second-order model (which is argued in §8.5 to provide a general description of non-Newtonian behaviour in sufficiently slow flow) leads to the following relation:

$$\left.\frac{d\eta_E}{d\dot{\epsilon}}\right|_{\dot{\epsilon}\to 0} = \tfrac{3}{2}(\Psi_1 + 2\Psi_2)|_{\dot{\gamma}\to 0}. \tag{5.13}$$

Available experimental evidence from shear-flow rheometry (cf. §4.2) would indicate that

$$\left.\begin{array}{l} \Psi_1 \geqslant 0, \quad \Psi_2 \leqslant 0, \\ 0 \leqslant |\Psi_2| < 0.2\Psi_1, \end{array}\right\} \tag{5.14}$$

so we expect

$$\left.\frac{d\eta_E}{d\dot{\epsilon}}\right|_{\dot{\epsilon}\to 0} > 0 \tag{5.15}$$

for non-Newtonian elastic liquids. This is important, since it indicates that the extensional viscosity $\eta_E$ must be an increasing function of $\dot{\epsilon}$ for very small values of $\dot{\epsilon}$, i.e. *initial tension-thickening* must be anticipated for all elastic liquids satisfying (5.14), whatever the response at higher values of $\dot{\epsilon}$ may be.

Bird (1982) has made the interesting observation that it is usually easier to calculate theoretically the *extensional* flow characteristics of molecular models than the corresponding *shear* flow functions. He has also provided a useful summary of the $\eta_E$ predictions for various molecular models of polymeric liquids. Many of them predict infinite extensional viscosities at a finite value of the extensional strain rate (cf. the predictions in Table 8.3). Partly to overcome this problem, Phan Thien and Tanner (1977) proposed a model which allows $\eta_E$ to pass through a maximum rather than take infinite values. For general and future interest, we show in Fig. 5.4 schematic $\eta_E$ and $\eta_{EP}$ curves computed for the so-called PTT model. Interestingly, although the initial values of $\eta_E$ and $\eta_{EP}$ are different at low strain rates, in agreement with eqns. (5.8) and (5.9), they are indistinguishable at high strain rates.

The general observation concerning the relative ease of carrying out theoretical work on the extensional flow characteristics of molecular models is also illustrated

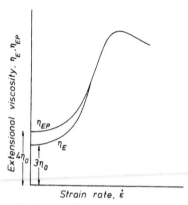

Fig. 5.4 Showing how the planar, $\eta_{EP}$, and uniaxial, $\eta_E$, extensional viscosities vary with strain rate $\dot{\epsilon}$ for the Phan-Thien–Tanner (PTT) model.

by the important work of Batchelor (1970, 1971) on suspensions of slender particles. He showed that the extensional viscosity for such systems can be very high, depending on the aspect ratio of the particles.

## 5.4 Experimental methods

### 5.4.1 General considerations

It is generally agreed that it is far more difficult to measure extensional viscosity than shear viscosity, this being especially so for mobile liquids. There is therefore a gulf between the strong desire to measure extensional viscosity and the likely expectation of its fulfilment.

Concerning experimentation, we remark that, in the case of stiff systems, the basic problem is not one of exposing the sample to a uniaxial extensional flow, but rather of maintaining it for a sufficient time for the stress (in a controlled strain-rate experiment) or the strain rate (in a controlled stress experiment) to reach a steady state, thus enabling the *steady* extensional viscosity $\eta_E$ to be determined. This is nowhere better illustrated than in the careful experimentation on the LDPE sample commonly referred to as IUPAC A (see, for example, Meissner et al. 1981, and Fig. 5.12). Extensive early work up to Hencky strains * of 5 or 6 seemed to indicate that an equilibrium had been reached, thus permitting the calculation of $\eta_E$. However, further experiments involving strains of up to 7 have indicated that the "equilibrium" was in fact a "turning point" and the ultimate equilibrium value of $\eta_E$, if it exists, must be lower than the original (overshoot) value (see, for example, Meissner et al. 1981, Meissner 1985). However, the new data do not point unambiguously to a new equilibrium value and Bird et al. (1987(a), p. 135) stress the

---

\* The Hencky strain $\epsilon$ is defined as $\ln(L/L_0)$ where $L$ is the final length of the sample whose original length is $L_0$.

general difficulty of reaching a steady state in extension for many polymer melts, questioning whether a meaningful extensional viscosity exists for some materials and conditions. Fortunately, a knowledge of the transient function $\bar{\eta}_E(t,\dot{\epsilon})$ may be sufficient (or at least very useful) in many practical applications (cf. Bird 1982, Laun and Schuch, 1988).

When it is realized that the Hencky strain of 7 reached in the Meissner experiments corresponds to stretching the sample to 1100 times its original length, the difficulties involved in extensional rheometry become self-evident.

The problems of determining the extensional viscosity of *mobile* liquids are even more acute, but they are of a different type from those experienced for stiff systems. With mobile liquids, severe difficulties arise in trying to achieve a continuous extensional flow field which approximates that given in eqs. (5.1). The most that one can hope for is to generate a flow which is dominated by extension and then to address the problem of how best to interpret the data in terms of material functions that are rheologically meaningful.

Fortunately, for many mobile elastic liquids, the extensional viscosity levels are so high (and potentially important) as to justify such an approach. This fact has spawned a number of extensional rheometers in recent years and most of them are able to capture the high extensional viscosities which are known to exist. The main outstanding problem is to assess critically the "viscosities" arising from the various methods and to see whether a concensus emerges. This is under active consideration (Walters 1988).

### 5.4.2 Homogeneous stretching method

The homogeneous stretching method, illustrated in Fig. 5.5, was the first to be used to determine the extensional viscosity.

A major (unavoidable) disadvantage of this method is that, in order to attain a constant extensional strain rate in the sample, the velocity of the movable block must vary exponentially with time. In principle this can now be accomplished very easily with the most recent electronic-control techniques, but the accelerating motion of the clamp places a severe constraint on the strain rates which can be attained, given the requirement that the motion must be sustained for a sufficient time for the stress (which is measured at either the stationary or the moving block) to reach a steady value.

For practical reasons the overall deformation is clearly restricted in the conventional stretching method of Fig. 5.5. The Meissner (1972) apparatus shown in Fig. 5.6 goes some way to overcoming this problem (see also Laun and Münstedt 1978). Instead of end loading, constant stretching is provided by two sets of toothed wheels

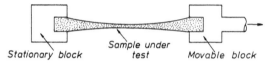

Fig. 5.5 Schematic diagram of the homogeneous stretching method.

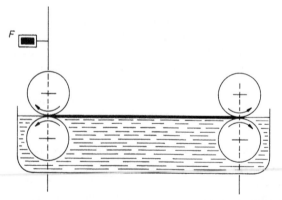

Fig. 5.6 Schematic diagram of the Meissner apparatus for attaining high strains (see, for example, Meissner 1972). Note that the stretched specimen is supported by a suitable liquid.

which rotate with constant angular velocity. The stress can be measured by the deflection of a spring F, associated with one pair of rollers.

Using this method on polymer melts, Meissner (see, for example, Meissner 1985) has been able to reach Hencky strains as high as 7. A further recent development has involved the use of a series of clamps in the form of a ring. In this way other modes of (multiaxial) extensional deformation can be generated (Meissner 1985).

The homogeneous stretching method is clearly restricted to high-viscosity systems.

### 5.4.3 Constant stress devices

The instruments shown schematically in Figs. 5.5 and 5.6 are of the constant strain-rate type. An alternative technique, first introduced by Cogswell (1968) and developed later by Münstedt (1975, 1979) utilizes a constant *stress*, which is brought about by applying a force to the movable block in Fig. 5.5, the force decreasing in proportion to the cross-sectional area of the extending specimen. Cogswell exployed this method using a cam to apply a programmed load, together with a convenient means of measuring the length of the sample as a function of time.

It is interesting to note that as a general rule the constant stress devices reach the steady-rate elongational flow regime at smaller total deformations than the constant strain-rate devices. For example, Laun and Schuch (1989) quote that a strain of 3.5 was required to reach equilibrium in a constant stress experiment on an LDPE melt whereas a strain of 4.5 was required in the comparable constant strain-rate experiment.

The constant stress devices are also clearly restricted to high-viscosity systems.

### 5.4.4 Spinning

It is self-evident that fibre spinning involves a significant extensional-flow component. At the same time, it is extremely difficult (if not impossible) to interpret

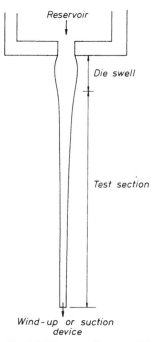

Reservoir

Die swell

Test section

Wind-up or suction
device

Fig. 5.7 Schematic diagram of the spin-line rheometer.

the data unambiguously in terms of the extensional viscosity $\eta_E$ defined in eqns. (5.2). The problem is that, although the flow may be steady in an Eulerian sense (in that the velocity at a fixed distance down the threadline does not vary with time) it is unlikely to be steady in a Lagrangian sense (since the strain rate experienced by a given fluid element will generally change as it moves along the threadline). Furthermore, even when the strain rate is constant over a *portion of the threadline* (so that a given fluid element is exposed to a constant strain rate for a limited period of time) that element may still "remember" conditions experienced in the reservoir in the case of highly elastic liquids. Certainly, there is ample evidence that a change of conditions in the spinnerette can significantly affect the response along the threadline under some conditions.

The fibre-spinning experiment (Fig. 5.7) is therefore a typical illustration of the dilemma facing rheologists who are interested in extensional-viscosity measurement. It is relatively easy to perform, the general kinematics can be determined with relative ease, and a suitable stress variable can be obtained from force measurements on the reservoir or the take-up device (see, for example, Hudson and Ferguson 1976, Jones et al. 1987). However, a consistent quantitative interpretation of the experimental data in terms of the extensional viscosity $\eta_E$ defined in eqns. (5.2) is not possible. One can certainly define *an* extensional viscosity by dividing the measured stress by (say) an average value of the strain rate (Jones et al. 1987). Given the difficulties encountered in measuring extensional viscosity and the scale of the viscoelastic response in such a flow, the proposed course of action can be justified and may be all that is required in many circumstances.

The spinning technique can be used for polymer melts (see, for example, Laun and Schuch 1989) and for low viscosity liquids (Jones et al. 1987). In the commercial spin-line rheometer (see, for example, Ferguson and El-Tawashi 1980) the wind-up device is a rotating drum. A variant of this for *very* mobile liquids is to use a suction device (Gupta and Sridhar 1984).

### 5.4.5 Lubricated flows

A schematic diagram of the lubricated-die rheometer is given in Fig. 5.8. The shape of the test section is so designed that the flow is equivalent to steady extensional flow *if there is perfect slip at the walls*. To facilitate this, lubricant streams of low-viscosity Newtonian liquids are employed. Pressure measurements provide the relevant stress input.

This would seem to be a convenient technique for eliminating the unwanted shearing induced by the rheometer walls, thus providing a flow close to the desired extensional flow. However, the interface boundary condition between the sample and the lubricant is dependent on the rheological properties of the sample. The interpretation of experimental data is therefore not without its problems and the technique itself is far from easy to use (see, for example, Winter et al. 1979, Williams and Williams 1985, Jones et al. 1987).

A similar technique has also been used by Winter (see, for example, Soskey and Winter 1985) to study *biaxial* extensional flows of polymer melts. In this case uniform discs of the sample are placed between two parallel circular plates, both of

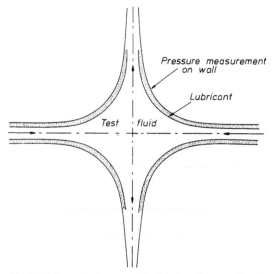

Fig. 5.8 Schematic diagram of the lubricated-converging-flow rheometer; the shape of the channel walls is chosen such that the flow is equivalent to a pure extensional flow if the low-viscosity lubricant streams are able to bring about "perfect slip" at the walls.

which are coated with a low-viscosity lubricant. The plates are then squeezed together. The relationship between applied load and rate of squeezing is interpreted in terms of the biaxial extensional viscosity $\eta_{EB}$. Again, data interpretation is difficult and the technique itself is not easy to use.

### 5.4.6 Contraction flows

The contraction-flow method of determining extensional properties can be applied equally well to polymer melts and to more mobile systems like dilute polymer solutions. Indeed, in the case of polymer melts, the so-called Bagley (1957) correction (or a suitable variant of it) must be used to interpret correctly *shear* viscosity data from a capillary rheometer (cf. §2.4.9). The Bagley correction can be immediately utilized to yield *extensional*-viscosity information on the melt, as we shall see. Specifically, in capillary rheometry where the test fluid is forced to flow under pressure from a barrel into a capillary of much smaller radius, one technique is to measure the pressure in the barrel for fixed capillary diameter $D$ and varying length $L$ (cf. Chapter 2 and Fig. 5.9). At a fixed flow rate, a plot of pressure as a function of $L/D$ provides sufficient information to facilitate shear-viscosity determination: in particular, the pressure drop for *fully developed* Poiseuille flow along a capillary of a given length can be determined.

Often, just two experiments are carried out in capillary rheometry: one for a capillary of a reasonable length ($L/D \geqslant 20$) and the other for so-called orifice flow (i.e. $L = 0$). It is this last experiment (and by implication the Bagley correction) which is of potential importance in the determination of extensional-viscosity characteristics.

In general, a contraction geometry is simply two capillaries of different diameters with an abrupt contraction between them. In some experiments, as we have indicated, flow through an orifice is an alternative.

Usually, but not always, the flow consists of a central core and a vortex region (see Fig. 5.10). The kinematics are determined by the flow rate and the shape of the central core region, whilst the relevant stress is obtained from the pressure drop required to force the test fluid through the contraction. There is no doubt that the

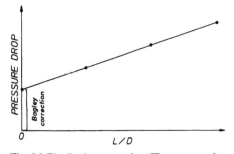

Fig. 5.9 The Bagley correction. The pressure drop (for a fixed flow rate and fixed capillary diameter) is measured for various values of capillary length.

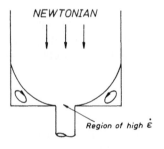

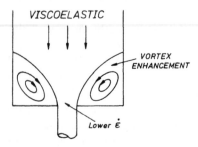

Fig. 5.10 Contraction flow. In the case of highly elastic liquids, vortex enhancement usually occurs.

contraction-flow experiment is relatively easy to perform and has many attractive features.

An approximate analysis of the contraction-flow problem has been developed by Binding (1988). This extends and reinterprets the early analysis of Cogswell (1972(a) and (b)). The Binding analysis is based on the assumption that the flow field is the one of least resistance; it includes both shear and extension in its formulation. The theory successfully predicts the phenomenon of vortex enhancement, which is often observed in axisymmetric contraction flows, and provides estimates of the extensional viscosity.

### 5.4.7 Open-syphon method

The open-syphon technique is shown schematically in Fig. 5.11. Fluid from the reservoir is drawn up through a nozzle by a vacuum pump and the nozzle is then raised above the level of the liquid in the reservoir. With some liquids the upward flow continues. This is the open-syphon effect. The flow rate and the dimensions of the fluid column yield the relevant kinematical information and the stress is provided by force measurements made at the top of the liquid column (see, for example, Astarita and Nicodemo 1970, Moan and Magueur 1988).

In general terms, the open-syphon technique suffers from the same general disadvantages as the spinning experiment, so far as data interpretation is concerned. However, the former may have advantages in the case of structured materials like gels. In the spinning experiment the structure is often changed by shear in the delivery pipe. In contrast, the test material in the open-syphon technique is in its

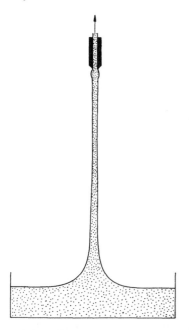

Fig. 5.11 The open-syphon technique for studying extensional flow. Liquid is sucked up from the reservoir into a tube, and the downward pull on the tube is measured.

rest state before being exposed to a sudden extension and one is therefore determining what is essentially a measure of the extensional properties of the virgin gel.

### 5.4.8 Other techniques

Numerous other techniques have been suggested for the study of the extensional behaviour of mobile elastic liquids. These include the so-called triple-jet technique (Oliver and Bragg 1974), the droplet techniques of Schümmer and Tebel (1983) and Jones and Rees (1982), the elongation of radial filaments on a rotating drum (Jones et al. 1986) and stagnation-point devices such as the opposing-jet techniques of Odell et al. (1985), Keller and Odell (1985) and Fuller et al. (1987).

## 5.5 Experimental results

Typical *transient* extensional-viscosity data for a polymer melt are given in Fig. 5.12 (cf. Meissner 1985, Bird et al. 1987(a)). It will be seen that as the strain rate $\dot{\epsilon}$ is increased, the experimental results depart from "Trouton behaviour" *, increasing abruptly with time. This is called "strain hardening". There then follows the maximum in $\bar{\eta}_E(t, \dot{\epsilon})$ already referred to in §5.4.1. In view of the extreme difficulty

---

* Trouton behaviour, in this context, is obtained from linear viscoelasticity theory.

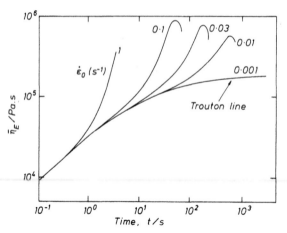

Fig. 5.12 Extensional viscosity growth $\bar{\eta}_E(t, \dot{\epsilon})$ as a function of time $t$ for a low-density polyethylene melt. 423 K (see, for example, Meissner 1985).

of obtaining data at relatively high strains, rheologists nowadays often simply quote the maximum values of $\bar{\eta}_E(t, \dot{\epsilon})$ (which were, of course, once thought to be the equilibrium values $\eta_E(\dot{\epsilon})$). When this is done, one obtains the type of result shown in Fig. 5.13 for four polymer melts. The data are consistent with the requirements of eqs. (5.8) and (5.15). There is a clear indication of a maximum in $\eta_E$ with strain rate for the polyethylenes.

   If the extensional viscosity and shear viscosity are plotted as functions of stress,

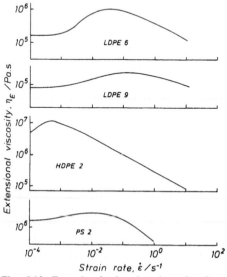

Fig. 5.13 Extensional viscosity data for four polymer melts (after Laun and Schuch 1989). LDPE—low-density polyethylene; HDPE—high-density polyethylene; PS—polystyrene.

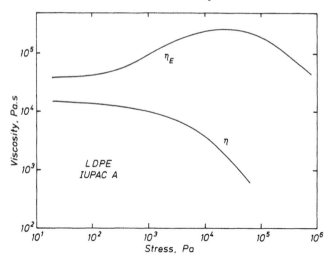

Fig. 5.14 Extensional viscosity and shear viscosity as functions of stress for the low-density polyethelyne designated IUPAC A. 423 K (cf. Fig. 5.12) (see, For example, Laun and Schuch 1989).

rather than strain rate, the response shown in Fig. 5.14 is obtained for a typical low-density polyethylene melt.

In Fig. 5.15 we show extensional-viscosity data obtained from a spin-line rheometer on a solution of polybutadiene in dekalin (Hudson and Ferguson 1976). Here there is a further and substantial increase in $\eta_E$ after the tension-thinning region. The data in Figs. 5.13 and 5.14 do not extend to sufficiently high values of $\dot{\epsilon}$ to indicate whether the ultimate tension-thickening trend occurs also for stiff polymeric systems.

We have already indicated that the accurate determination of $\eta_E$ for mobile elastic liquids is very difficult, perhaps impossible, but the evidence to hand indicates that the extensional viscosities which have been measured can be very high indeed. For example, in Figs. 5.16 and 5.17 we show shear- and extensional-viscosity

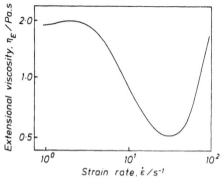

Fig. 5.15 Extensional viscosity curve determined with a commercial spin-line rheometer for a 6.44% solution of polybutadiene in dekalin (cf. Hudson and Ferguson 1976).

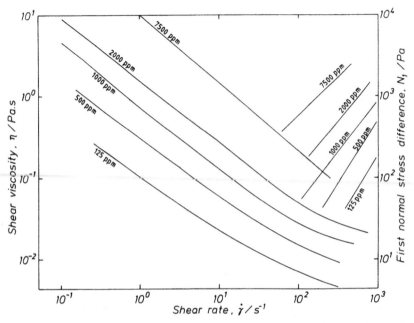

Fig. 5.16 Viscometric data for aqueous solutions of polyacrylamide (1175 grade) (Walters and Jones 1988). Note that viscosity decreases and normal stress increases with shear rate.

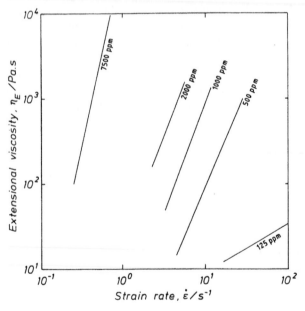

Fig. 5.17 Extensional viscosity data obtained from a spin-line rheometer for the aqueous polyacrylamide solutions of Fig. 5.16 (Walters and Jones 1988). Note that whereas shear viscosity decreased with shear rate, extensional viscosity increases with extensional rate.

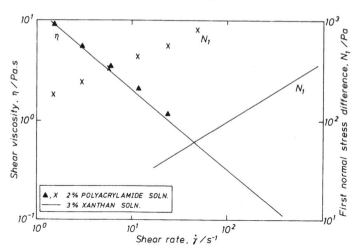

Fig. 5.18 Viscometric data for a 2% aqueous solution of polyacrylamide (E10 grade) and a 3% aqueous solution of Xanthan gum. Note the very different values of $N_1$ for solutions with almost the same viscosities.

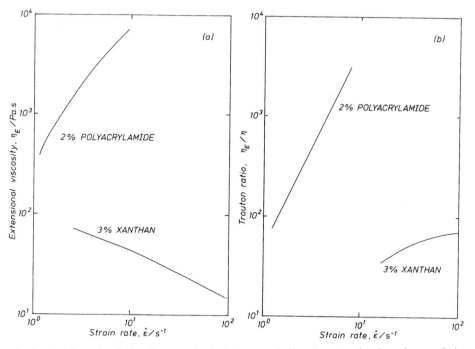

Fig. 5.19 (a) Extensional viscosity data obtained from a spin-line rheometer for the polymer solutions investigated in shear flow in Fig. 5.18; (b) Trouton ratios obtained from Figs. 5.18 and 5.19(a). Note that although the Xanthan gum solution is tension-thinning (Fig. 5.19(a)), the associated Trouton ratios increase with strain rate and are still significantly in excess of the inelastic value of 3.

data for a series of very dilute aqueous solutions of a very high molecular-weight polyacrylamide designated 1175 (Walters and Jones 1988). It is not difficult to deduce that the Trouton ratios are as high as $10^4$.

We now refer to the rheometrical behaviour of two polymer solutions with almost identical shear viscosity behaviour (see Figs. 5.18 and 5.19). One is a 2% aqueous solution of polyacrylamide (E10 grade) and the other a 3% aqueous solution of Xanthan gum. We note that polyacrylamide is a more flexible polymer than Xanthan gum which is rod-like. The corresponding extensional-viscosity data obtained from a spin-line rheometer show that the polyacrylamide solution is strongly tension-thickening, whereas the Xanthan gum solution is tension-thinning over the

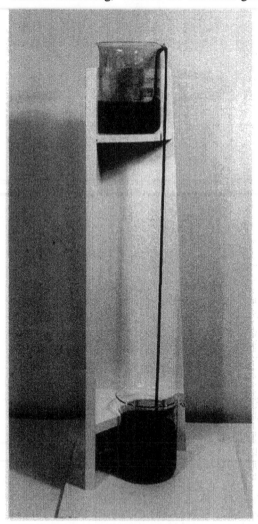

Fig. 5.20 An illustration of the open-syphon effect for a 0.75% aqueous solution of polyethylene oxide.

range studied. However, when the Trouton ratios are calculated on the basis of eqn. (5.12), we see that even the Xanthan gum solution has $T_R$ values that are significantly higher than the Newtonian value of 3 over most of the range.

Finally, we note that available evidence would indicate that the Trouton ratios for non-polymeric colloidal liquids are much lower than those for polymeric liquids showing similar behaviour in small-amplitude oscillatory-shear flow.

## 5.6 Some demonstrations of high extensional viscosity behaviour

We conclude this chapter by referring to some easily reproduced situations in which the dramatic effects of high Trouton ratios are clearly in evidence. Some of these can be readily performed in the laboratory with quite standard equipment.

We have already referred to the open-syphon technique for measuring extensional-viscosity (cf. §5.4.7). We remark that the use of a 0.75% aqueous solution of polyethylene oxide WSR 301 grade (or similar polymer solution) will enable the experimenter to operate a conventional syphon several centimetres above the level of the reservoir liquid.

The open-syphon effect is even more dramatically demonstrated when the polymer solution is transferred from one full container to a lower empty container. All that is normally required to (almost) empty the container is to start the flow by slightly tilting the top container. The initial flow will be sufficient to empty the bulk of the liquid from the top container (see Fig. 5.20). The open-syphon phenomena can be directly attributed to the very high Trouton ratios exhibited by the polymer solution. These and other dramatic demonstrations of high extensional-viscosity behaviour are illustrated in the film *"Non-Newtonian Fluids"* produced by Walters and Broadbent (1980) *.

Our final example of visual extensional-viscosity phenomena is provided by flow past cylindrical obstructions placed asymmetrically in a parallel channel (cf. Walters and Jones 1988). The flow is basically two-dimensional and may be considered to be made up of narrow channels and wide channels formed by the offset positioning of the circular-cylinder barriers (Fig. 5.21).

The behaviour of the Newtonian liquid is unspectacular with as much liquid going through the narrow channels as one would expect. This behaviour may be contrasted with that for the relatively inelastic Xanthan gum solution. In this case, a substantial flow finds its way through the narrow channels, clearly on account of the shear-thinning viscosity.

Finally, we note the qualitative difference in the behaviour of the highly elastic shear-thinning polyacrylamide solution. In this case, extensional viscosity consider-

---

* The film is available in Video or 35mm form from the Department of Mathematics, University College of Wales, Aberystwyth, UK.

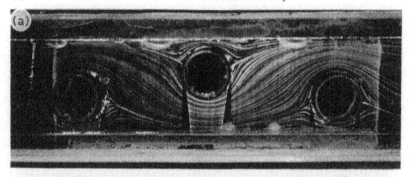

Fig. 5.21 Flow visualization pictures for: (a) a Newtonian liquid; (b) a Xanthan gum solution; (c) a polyacrylamide solution. They show the dominant effects of shear-thinning in the Xanthan gum solution and tension-thickening in the polyacrylamide solution (see, for example, Walters and Jones 1988).

ations are all important and virtually no liquid finds it way into the narrow channels. There is a relatively fast-moving stream in the wide channels and virtually stagnant regions elsewhere (cf. the four-roll mill situation in Fig. 5.3).

CHAPTER 6

# RHEOLOGY OF POLYMERIC LIQUIDS

## 6.1 Introduction

The rheological literature is dominated by discussions of polymer rheology. The reasons for this are not difficult to determine; the subject is extremely important industrially and much money and many resources are expended in carrying out the relevant research. Also, polymeric liquids exhibit a wide range of rheological phenomena and can often be tailor-made to facilitate fundamental rheological research. Indeed, it should not have escaped the reader that most of the examples used in earlier chapters to illustrate various rheological phenomena were obtained with polymeric liquids.

The domination of polymer rheology is also reflected in the number of books either devoted to the subject or strongly influenced by it. We shall have cause to mention some of these in the course of this chapter, but at this point we refer especially to the texts of Bird et al. (1987(a) and (b)) which contain a thorough study of "the dynamics of polymeric liquids" and together are recommended as an (almost) encyclopaedic treatise on the subject.

The fact that there are so many detailed and expert texts on polymer rheology and that the subject itself is very broad in scope suggests that, in the present *introductory* text, we should do no more than attempt an overview of the subject, pointing the interested reader to the preferred texts on matters of detail.

## 6.2 General behaviour

The generic term "polymeric liquids" can be viewed as including a spectrum of possibilities, ranging from mobile systems like very dilute polymer solutions, through the concentrated solution regime to stiff systems like polymer melts. All the rheological properties introduced in earlier chapters can be demonstrated to occur in suitably chosen polymer systems. The wide diversity of observed phenomena is attributable to the long chain molecules, which are a unique characteristic of polymers. The length of the chain is the main factor determining the rheology, although many other factors are also influential, as we shall indicate.

A long chain will occupy a great deal of space compared to its atomic dimensions. The possibility of polymer molecules linking together, either temporarily by intermolecular forces or more permanently by chemical cross linking (as in the

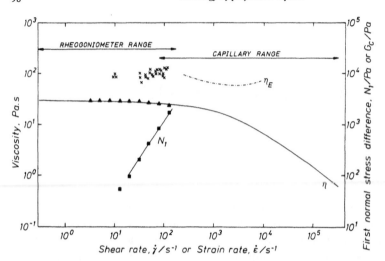

Fig. 6.1 Steady shear and extensional viscosity data for a concentrated solution of a semi-rigid polymer of modest molecular weight. 25° C. Full lines (————) steady shear data obtained with a rheogoniometer and a capillary rheometer (points not shown). Crosses (× × ×) and broken line (–·–·–) extensional viscosity from the spin-line rheometer and contraction flow respectively. Triangles (▲▲▲) predictions based on the Cox–Merz rule for viscosity. Squares (■■■) $N_1$ derived from $|G_c|$ as described in §6.10.

vulcanization of rubber) increases still further the space over which the influence of an individual molecule is felt. If the polymer chain is long enough, the intermolecular association known as *entanglement* occurs. The entangled polymer, whether in the molten state or in non-dilute solution, gives rise to the effects of high elasticity, such as normal stresses and high extensional viscosity.

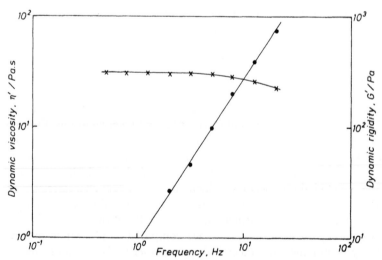

Fig. 6.2 Oscillatory data for the polymer solution of Fig. 6.1 obtained with a rheogoniometer. 25° C.

The vast majority of polymeric liquids exhibit shear-thinning in a steady simple-shear flow (cf. §2.3.2). Figure 6.1 contains a typical example, for a concentrated solution (30%) of a stiff-chain polymer of moderate molecular weight ($\sim$ 20,000). Such behaviour may be viewed as being of modest proportions and it is certainly possible to generate more severe shear-thinning by, for example, increasing the molecular weight of the polymer. However, the data in Fig. 6.1 clearly demonstrate the existence of a first (lower) Newtonian region followed by the shear-thinning zone. Figure 6.2 contains the corresponding dynamic data obtained from a small-amplitude oscillatory-shear flow (cf. §3.5). When the Cox–Merz rule * is applied to the dynamic data, we see from Fig. 6.1 that the shear viscosity behaviour is predicted very well over the available range of the data.

In some circumstances, because of experimental limitations, it is not possible to reach the second (upper) Newtonian region, and it is even more difficult to reach sufficiently high shear rates to check on the existence or otherwise of the ultimate (shear-thickening) upturn in viscosity mentioned in §2.3.3. However, recent studies have demonstrated that shear-thickening, sometimes accompanied by antithixotropy, can be obtained in dilute polymer solutions beyond a critical set of conditions (see, for example, Jackson et al. 1984).

Figure 6.1. also contains first normal stress data for the polymer solution. The normal stress level is also modest and it is certainly possible to obtain much higher normal stresses in polymeric liquids. $N_2$ data are not available for this polymer solution, but from the discussion of §4.2 we would anticipate $N_2$ to be negative and much smaller than $N_1$.

Extensional-viscosity results for the polymer solution are also included in Fig. 6.1. These were obtained from the spin-line rheometer and a contraction flow device (cf. §5.4.6). $\eta_E$ is seen to be relatively constant over the range of the experiments. Since $\eta(\dot{\gamma})$ falls as $\dot{\gamma}$ increases, it is evident that the resulting Trouton ratios are greater than 3 over much of the experimental range.

The extensional-viscosity behaviour of polymeric liquids is maybe the one area where it is possible to distinguish qualitatively between the behaviour of dilute polymer solutions and polymer melts. We have already alluded to this in §5.5, where we showed that the general viscosity behaviour of dilute polymer solutions and polymer melts was similar to that shown schematically in Fig. 6.3. In the former, $\eta_E$ rises abruptly with strain rate $\dot{\epsilon}$ after some critical strain rate $\dot{\epsilon}_c$; $\eta_E$ can reach very high values indeed. In constrast, for a polymer melt, $\eta_E$ is usually a weak function of $\dot{\epsilon}$, by which we mean that the magnitude of $\eta_E$ does not change very much as $\dot{\epsilon}$ is varied. Note, however, the *indication* that, even in the case of polymer melts, $\eta_E$ may ultimately rise sharply with increasing $\dot{\epsilon}$. Arriving at a firm conclusion on this matter is hindered by the difficulty of obtaining consistent experimental data at

---

* The empirical Cox–Merz (1958) rule states that the shear viscosity $\eta$ should be the same function of shear rate $\dot{\gamma}$ as $|\eta^*|$ is of frequency $\omega$, where $|\eta^*| = [(\eta')^2 + (G'/\omega)^2]^{1/2}$, $\eta'$ and $G'$ being the dynamic viscosity and dynamic rigidity, respectively (see Chapter 3).

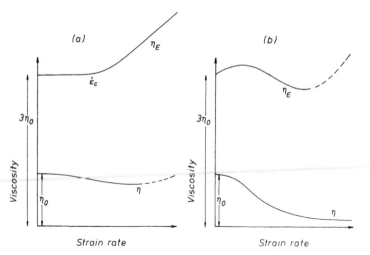

Fig. 6.3 Schematic representation of typical shear viscosity $\eta$ and extensional viscosity $\eta_E$ behaviour for: (a) A dilute polymer solution; (b) A concentrated polymer solution or polymer melt.

sufficiently high strain rates, but recent molecular theories are not inconsistent with such a trend (Marrucci 1988).

Note that for all polymeric liquids, the Trouton ratio is expected to be always greater than or equal to 3, with the value 3 only attained at vanishingly small strain rates (cf. Chapter 5).

We make passing reference to the one type of highly elastic polymer solution which does not exhibit measurable shear thinning. We refer to the so-called Boger (1977(a)) fluids, which are very dilute solutions of a high molecular-weight polymer in a solvent with a *high* viscosity $\eta_s$. The dissolved polymer only contributes a small proportion (say 5%) to the final zero-shear viscosity $\eta_0$ of the polymer solution. This means that the viscosity is essentially confined to lie between $\eta_0$ and $\eta_s$, so that the shear-thinning is constrained to be at most 5% and, in practical terms, it appears from conventional rheometry that the polymer solution has a constant viscosity. At the same time, the presence of the high molecular-weight polymer, even at very low concentration levels, can result in substantial normal stresses and high extensional viscosities, and the polymer solution may be (loosely) regarded as being a highly elastic constant-viscosity liquid *. Examples of such a liquid include very dilute solutions ($\sim 0.1\%$) of polyacrylamide in a maltose syrup/water base and very dilute solutions (0.1%) of polyisobutylene in a mixture of kerosene and low molecular-weight polybutene (see, for example, Prilutski et al. 1983). Boger fluids have proved to be popular test fluids in the study of viscoelastic effects in complex flows (see, for example, Binding et al. 1987).

---

* Available experimental evidence would suggest that $N_2 \approx 0$ for Boger fluids (see Fig. 4.5 and Keentok et al. 1980).

## 6.3 Effect of temperature on polymer rheology

It is now instructive to consider the changes which can occur in a given polymer system when a variable parameter is scanned over a range. As a first example, we take a thermoplastic polymer and change the temperature (see also §2.2.2 and cf. Tanner 1985, Chapter 9).

At temperatures well above the melting point $T_m$, the polymer is a liquid with a measurable (shear-thinning) viscosity. It is possible to observe normal stresses in simple shear flow and also relatively high extensional viscosities, which indicate that the melt is viscoelastic.

As the temperature is reduced the viscosity increases rapidly and the elasticity becomes more evident. In this state the melt displays a pronounced elastic recovery from any deformation. It is this ability to recover from large deformations that justifies the description of the behaviour as 'highly elastic'.

At still lower temperatures, some polymers crystallize and the freezing point is a first-order transition which involves latent heat. Polyethylene is an example of a crystallizing polymer. The semi-crystalline solid polymers have a shear modulus of about 1 GPa. This value is about $10^4$ times higher than the modulus of a typical unhardened rubber but is lower than that for a metal. The polymer in this state can in general undergo larger deformations than metals without fracturing.

Other polymers do not crystallize but continue to increase in viscosity, eventually to form a glass. Polystyrene is an example of a non-crystallizing polymer. The glass transition is not a well-defined change of state and depends on the method of measurement. It is not accompanied by a latent-heat change. Some liquid-state theories associate the glass transition temperature $T_g$ with the attainment of a particular value of the free volume. Since free volume and viscosity are closely related at high viscosities, $T_g$ has also been associated with the attainment of a

TABLE 6.1
Transition temperatures and operating temperatures for some common polymers

|  | $T_m$ (°C) | $T_g$ (°C) | Normal melt-processing temperatures (°C) |
|---|---|---|---|
| High density polyethylene (HDPE) Low density polyethylene (LDPE) | 140 | −100 | 160–240 |
| Isotactic polypropylene (PP) | 165 | −15 | 180–240 |
| Polyethylene terephthalate (PET) | 265 | 70 | 275–290 |
| Nylon-66 | 265 | 40 | 275–290 |
| Polystyrene | n/a | 100 | 180–240 |

certain viscosity, which has a value of about $10^{12}$ Pa.s. Since polymers do not crystallize readily, it is reasonably easy to prevent crystallization by rapid cooling or by introducing a degree of chain branching into the molecules.

Table 6.1 contains $T_m$ and $T_g$ values for some common polymers (from White 1980; see also Tanner 1985 p. 351).

## 6.4. Effect of molecular weight on polymer rheology

The second example illustrating general rheological behaviour of polymer systems is provided by a series of polymers at a given temperature but covering a range of molecular weights.

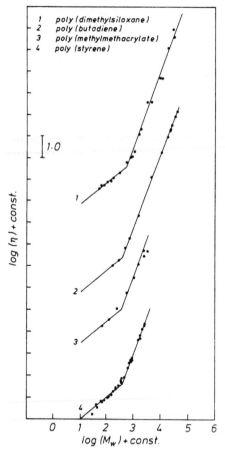

Fig. 6.4 Variation of zero-shear melt viscosity with molecular weight. The data have been scaled using constant factors in order that the change in slope should occur at the same position and the curves should be separated. To the left of the transition the slope is 1.0, whilst to the right the slope is 3.4 (from Ferry 1980, p. 244).

The monomer, and molecules containing just a few monomer units, are indistinguishable (rheologically) from small-molecule liquids, but at and above a certain length, the molecules become polymeric. In very high-frequency oscillatory experiments, Lamb and his coworkers (see, for example, Gray et al. 1977) found no evidence of polymeric modes of motion in chains of polystyrene and other monomers less than *eleven* monomer units long. In samples of greater chain length than this, they observed polymeric behaviour.

The zero-shear viscosity $\eta_0$ of the melt increases (initially) as $M^{1.0}$, where $M$ is the molecular weight. With a further increase in $M$, $\eta_0$ varies as $M^{3.4}$. The change in the slope of the $(\eta_0, M)$ curve is reasonably sudden and can be identified with a critical molecular weight denoted by $M_c$. Some representative graphs taken from the extensive work of Berry and Fox (1968) and quoted by Ferry (1980, p. 244) are shown in Fig. 6.4.

The melts in the region above the critical molecular weight are highly elastic. It is thought that the change in $M$-dependence is the result of the formation of *entanglements* between molecules. Entanglements are not merely the overlapping of adjacent molecules, since overlapping occurs at quite low molecular weight. Rather, entanglements are strong couplings, whose details are largely unresolved, but which act in a localized manner like chemical cross-links between molecules.

## 6.5 Effect of concentration of the rheology of polymer solutions

We now consider the example of a polymer in solution. The polymer and solvent chosen are such that, at very low concentrations, there is no strong interaction between polymer molecules. In this region, physical properties change in direct proportion to the concentration $c$. As the concentration is increased, the increment in viscosity over that of the solvent increases at a faster rate. Ultimately there may be a change to a regime in which the viscosity varies as $c^3$ or even higher powers. The example given in Fig. 6.5 shows an increase in slope from 1.5 to 4. The transition is associated with the formation of entanglements, as happens in a melt above $M_c$. It follows that the polymer in solution must be capable of forming entanglements, hence the molecular weight must be higher than $M_c$ for the strong concentration-dependence to occur. In this region the solution properties are very similar to those of the melt, namely strongly viscoelastic.

A useful parameter which has been found helpful when making comparisons between different polymers in solution is the so-called "reduced concentration". This is the product of concentration and 'intrinsic viscosity', $c[\eta]$, and is therefore dimensionless. * Elastic effects achieve prominence when $c[\eta]$ attains a value in the range 5 to 10, and develop very rapidly with further increases in the parameter.

---

* Note: in polymer rheology $[\eta]$ usually has the dimensions of volume per unit mass (i.e. a reciprocal concentration); whereas in suspension rheology, since concentration is expressed as a volume per unit volume, $[\eta]$ is dimensionless (see §7.2.3). Of course, the product $c[\eta]$ is always dimensionless.

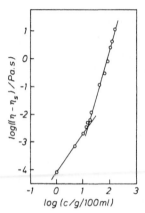

Fig. 6.5 Zero-shear viscosity of solutions of polystyrene in benzene showing the change from a 1.5-power dependence on concentration to a 4-power dependence. $\eta_s$ is the solvent viscosity (from Vinogradov and Malkin 1980, p. 187).

It should be understood that the three examples discussed in §6.3 to §6.5 are of specific rheological behaviour displayed by typical polymer systems; there are a number of variants. Thus, the details can be strongly influenced by such factors as molecular-weight distribution, chain branching, polarity, copolymerization, polymer mixtures (blends), suspended solid particles, solvent–polymer interactions and chain rigidity (see, for example, Brydson 1981, Vinogradov and Malkin 1980, Cogswell 1981, Nielsen 1977, White 1980, Laun and Schuch 1989). It is not appropriate for us to go into these in any detail, but two variants are worthy of special mention. These are polymer gels and polymeric liquid crystals.

## 6.6 Polymer gels

A polymer gel is generally a solution in which the chains are cross-linked by a more permanent means than mere physical entanglements. The cross-links may be chemical bonds of the type used to vulcanize or harden rubbers. Alternatively, they may be crystalline regions linked by chains which pass through more than one of these regions. The latter process is the same as that which occurs in solid semi-crystalline polymers. A rather different form of gel structure is obtained by adding high concentrations (say 20% v/v or more) of small solid particles such as carbon black to a polymer melt or concentrated solution. In such systems, the structure is formed by chains of the particles, although the formation of polymer bridges by adsorption on adjacent particles is an additional possibility.

A gel can be formed in solution in two ways: either by contacting the cross-linked solid with a suitable solvent, which is then taken up by the solid by the process of swelling, or by cross-linking molecules already in solution. Many natural products, such as gelatine, form gels of the latter type.

Gels may be regarded as 'soft solids' with a wide range of elastic modulus depending on the extent of cross-linking.

It is of interest that the particle-filled polymer gel, although having a high shear viscosity at low shear rates may have a relatively low extensional viscosity.

The book by Glicksman (1969) contains useful background information on polymer gels.

## 6.7 Liquid crystal polymers

Liquid crystals are formed in solutions of polymers which have a rigid backbone. Rigidity is conferred by introducing bulky groups such as benzene rings into the backbone structure. As the molecular weight is increased, chain entanglement begins at a critical value, as with flexible polymers. However, the concentration relationship for polymers above $M_c$ is very different. Initially, the viscosity increases rather rapidly with concentration beyond the entanglement point, but there then occurs a second critical point beyond which the viscosity *decreases* markedly. The phenomenon is shown in Fig. 6.6. The transition is a sharp one. It signals the formation of a strongly orientated polymer structure. Since orientation can be induced by shear, the position of the transition moves to lower concentrations as the shear rate increases.

Liquid crystals are also formed in melts of rigid-backbone polymers. The rheological behaviour of the liquid crystalline state is dependent on the orientation vector and its relation to the flow direction. This has led to the prediction of strange effects and may account for the observation of a negative first normal stress difference and zero, or negative, die swell in some liquid crystalline polymer melts, which in other respects are highly elastic.

For further information on this class of materials, the book edited by Chapoy (1985) is recommended.

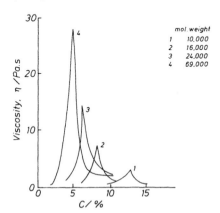

Fig. 6.6 Dependence of viscosity on concentration for solutions of poly(p-benzamide) in dimethyl acetamide at 20 °C. The liquid-crystalline state occurs to the right of the maxima in the curves (from Vinogradov and Malkin 1980, p. 196).

## 6.8 Molecular theories

### 6.8.1 Basic concepts

Viscoelastic phenomena in a polymeric liquid are due, primarily, to *intramolecular* forces which arise from the orientation of chemical-bond vectors in the polymer chains and, in particular, from changes in orientation caused by the deformation of the liquid. A molecule possesses a minimum-energy state (or rest state), in which the bond vectors are distributed in essentially a random configuration, and elastic recovery is a consequence of the return to this state. Therefore, the kinetic theory of rubber elasticity (see, for example, Treloar 1975) is basic to the physics of polymeric liquids. The presence of other molecules, whether polymer or solvent, delays the re-orientation process and gives rise to the viscous component of the rheological effect.

Starting from these fundamental ideas, molecular theories of polymer liquids follow two main branches. One branch is entitled *bead–spring* and the other *network*. They will be considered separately, although historically they were developed concurrently and are still being improved. For detailed treatments the reader is referred to Tanner (1985) and especially Doi and Edwards (1986) and Bird et al. (1987(b)).

### 6.8.2 Bead–spring models: the Rouse–Zimm linear models

In bead–spring models, viscous behaviour is specifically introduced by considering the molecule to be acted on by frictional drag according to Stokes' equation (see, for example, Bird et al. 1987(b)) or more elaborate forms of it. The bead–spring models, of which there are many, consider the sites of fluid friction to be represented by small spheres, or beads, which are connected by a length of polymer chain, which is itself considered to be frictionless. Figure 6.7 illustrates how the beads are imagined to be distributed along a chain. The chains between beads are equal in length and are sufficiently long for them to obey Gaussian statistics and thus for the chain elements to be entropic springs.

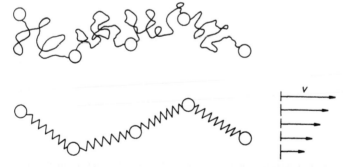

Fig. 6.7 The necklace model. Equal lengths of Gaussian chain (above) are represented by equal springs (below) joined by equal-sized beads which account for the drag experienced in the velocity field.

Two cases have been considered: the *dumbbell* models in which two beads are placed at the extremities of one spring and the more realistic but more complicated *necklace* models, in which (N + 1) beads are connected by N equal lengths of spring. The simple Stokes-drag law was used by Rouse (1953) in a notable paper on the necklace model. Zimm (1956) extended the theory to account for so-called hydrodynamic interaction.

The Rouse–Zimm theories apply to conditions of small-amplitude oscillatory shear flow and hence to linear viscoelastic behaviour (cf. Chapter 3). Specifically, the Rouse–Zimm treatment leads to linear viscoelastic models which are equivalent to distributed Maxwell elements, (see Fig. 3.7 (a)) with an interpretation of the model parameters in terms of molecular characteristics.

In passing, we remark that the constant-viscosity restriction implied in the linear theories of Rouse and Zimm is relaxed in the Bueche (1954) treatment. In this, shear thinning is predicted in steady simple shear flow. Specifically, $\eta$ is predicted to vary as $\dot{\gamma}^{-0.5}$. This rate of shear-thinning is reasonable for moderately concentrated polymer solutions but is low compared with most experimental results for concentrated solutions and melts.

### 6.8.3 The Giesekus–Bird non-linear models

We turn now to the advanced kinetic theories of principally Giesekus, and Bird and coworkers (cf. Giesekus 1985, Bird 1985, Bird et al. 1987(b)). It was the intention of these researchers to develop models which predict non-linear rheological effects, not just the linear ones as in the Rouse–Zimm treatment. In order to do this, it was necessary to use the dumbbell approximation (see Fig. 6.8). At first sight it seems to be a retrograde step to adopt the dumbbell approximation when the bead–spring necklace is intuitively a much better model of a real polymer chain. A major reason for making this step, and for its successful outcome, is that the longer relaxation-time processes are more influential than the short ones in conferring viscoelasticity, and the longest such process corresponds to the dumbbell.

The relevant theories lead to constitutive equations of the upper convected Maxwell and Oldroyd B type (cf. §8.6). Specifically, the steady shear viscosity $\eta$ is a constant, the first normal stress coefficient $\Psi_1$ is a positive constant and the second normal stress coefficient $\Psi_2$ is zero (cf. Chapter 4, eqs. 4.2 and 4.3).

Numerous developments of the basic dumbbell model have been made and these are discussed in detail by Bird et al. (1987(b)). Typical of these is the so-called

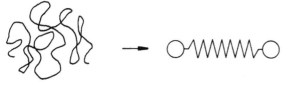

Fig. 6.8 The dumbbell approximation. The elastic behaviour is represented by a spring and the sites of the frictional drag are located at the ends of the spring.

FENE-dumbbell model * which predicts shear-thinning and a first normal stress coefficient which decays with shear rate. It also predicts stress overshoot in the start-up of shear flow.

### 6.8.4 Network models

The essential difference between the kinetic theory of rubber-like elasticity and the network models of polymer solutions and melts is that the permanent cross-links between molecules in the rubber are replaced by transient junctions in the network. The concept of transience is that, at any instant, junctions are sufficiently "permanent" for the network to behave like a rubber but they break after a short lifetime and reform elsewhere, hence the system is capable of flow. Since the total concentration of junctions always remains constant, the elastic properties are constant. The original idea of the transient network was published by Green and Tobolsky (1946). It was extensively developed by Lodge (1956, 1964) and is often associated with his name.

In a steady simple-shear flow, the Lodge rubber-like liquid predicts a constant viscosity $\eta$, a positive (constant) $\Psi_1$ and zero $\Psi_2$. It seems to be a coincidence that the Lodge theory leads to very similar predictions to the simpler versions of the Giesekus–Bird theories, especially since the molecular concepts are quite different in the two cases.

Numerous attempts have been made to modify the network theory by employing a more realistic molecular modelling and sometimes by adding further ad hoc assumptions (see, for example, Kaye 1966, Tanner 1969, Acierno et al. 1976, Johnson and Segalman 1977, Phan-Thien and Tanner 1977, Giesekus 1982, Leonov 1987). These attempts have been reasonably successful and it is now possible to predict, amongst other things, a second normal stress coefficient $\Psi_2$ of the correct form (see, for example, the predictions for the Johnson–Segalman and Phan-Thien–Tanner models in Chapter 8).

### 6.8.5 Reptation models

The entanglement theory of rubber elasticity of Edwards (1967) introduced a new physical concept which has had an influential impact on later developments in the molecular theory of polymer systems. The concept is of an equivalence between a chain constrained by entanglements, and/or cross-links, and a relatively free chain constrained in a *tube* and unable to escape through its sides. Figure 6.9 represents the equivalence pictorially. A related concept for uncrosslinked polymers introduced by de Gennes (1971) has revolutionized ideas on the mechanism of permanent deformation, or flow, of the entangled molecules. This concept is that the only way a molecule can escape from its tube, and hence allow flow to occur, is by diffusion along the tube. It is a wriggling, snake-like motion and the process has the aptly descriptive name of *reptation*. The detailed application of these concepts to polymer systems has been developed by de Gennes (1979), Doi and Edwards (1986) and

---

\* FENE stands for "finitely extensible, non-linear elastic".

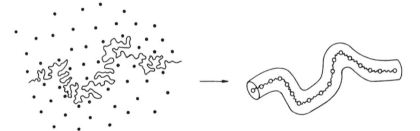

Fig. 6.9 the Doi–Edwards–de Gennes tube concept can be visualised by first placing the molecule on a plane (this page). Neighbouring molecules will intersect this plane at the points (see left-hand part of the figure). The nearest points define a tube which prevents major penetration by the molecule (see right-hand part).

Curtiss and Bird (see Bird et al. (1987(b)). The Doi–Edwards theory (with the additional "independent-alignment" assumption) leads to a constitutive equation similar to the so-called KBKZ model (see Chapter 8, eqn. 8.35). However, this statement simply indicates the broad general form of the constitutive equations arising from the simpler reptation theories and should be read in conjunction with the recent papers of Marrucci who has shown the limitations of the independent alignment assumption (Marrucci and Grizzuti 1986, Marrucci 1986).

There is no doubt that substantial progress has been made in deriving constitutive equations for polymeric liquids using 'microrheological' concepts. The interested reader is referred to the detailed texts of Tanner (1985 Chapter 5), Doi and Edwards (1986) and Bird et al. (1987(b)) for fuller details.

## 6.9 The method of reduced variables

In many cases it is important to study the full range of rheological properties for varying temperature and pressure (and in the case of polymer solutions, concentration as well). Such a study is facilitated by a suitable "normalization" procedure. The aim of such a procedure is to plot experimental results on a pair of axes using normalized variables, such that a unique curve is obtained whatever the temperature or pressure (or concentration). The so-called "method of reduced variables" is one way of accomplishing this. The method is fully described by Ferry (1980 p. 266), one of its originators, and it will be sufficient here to refer to the 'time–temperature superposition' principle only, although the additional variables, pressure and concentration, are also accommodated in the full method.

The method is basically empirical but is guided by reference to molecular theory. Time–temperature superposition depends on the principle that all the relaxation times in a spectrum have the same temperature dependence provided there is no phase change of the polymer within the temperature range considered. Hence

$$\tau_T = a_T \tau_0, \qquad\qquad (6.1)$$

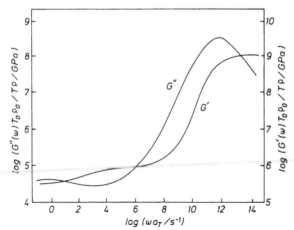

Fig. 6.10 Results of the application of time–temperature superposition to data for an unvulcanized rubber. Reference temperature $T_0 = 298K$. Shift factors $a_T$ were determined empirically by horizontal adjustment of individual curves (from Whorlow 1980, p. 417). Note that $\rho$ is the density.

where suffix $T$ refers to temperature $T$ and suffix 0 refers to the reference temperature $T_0$.

An example where the empirical shift process has been used for an unvulcanized rubber is shown in Figs. 6.10 and 6.11. The curves of $G'$ and $G''$ obtained at different temperatures (and not shown here) have been successfully reduced to two single composite curves by introducing a common temperature of 298K. The procedure in this case has resulted in values of $G'$ and $G''$ covering some 16 decades of frequency at 298K.

Empirically it is found that

$$\ln a_T = -c_1(T - T_s)/(c_2 + T - T_s), \tag{6.2}$$

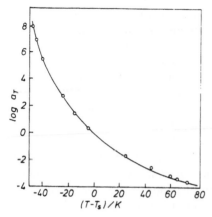

Fig. 6.11 Variation of shift factor $a_T$ for the rubber of Fig. 6.10 compared with the curve (solid line) predicted by the WLF equation ($T_s = 248K$, $c_1 = 8.86$ and $c_2 = 101.6K$).

in which $c_1$, $c_2$ and $T_s$ (a reference temperature) are constants. This form of the $a_T$ function has become known as the Williams–Landel–Ferry or WLF equation.

A test of the WLF equation is given in Fig. 6.11 which contains a graph of the shift factors used to produce Fig. 6.10.

For measurements of shear stress and normal stress in steady shear flow, the same principles apply with the reduced shear rate becoming $a_T\dot\gamma$.

In the case of a given polymer solution at varying concentration and temperature, it is often possible to construct a master curve by plotting $\eta/\eta_0$ against $\eta_0\dot\gamma$, where $\eta_0$ is the relevant zero-shear viscosity. Examples of such a procedure are provided by Vinogradov and Malkin (1980, p. 207) who also discuss possible extensions and limitations of the procedure.

## 6.10 Empirical relations between rheological functions

We have already referred to the Cox–Merz rule which provides a useful empirical correlation between the steady shear viscosity function $\eta(\dot\gamma)$ and the dynamic data ($\eta'(\omega)$ and $G'(\omega)$) in the case of polymeric liquids (see §4.5). A corresponding correlation between the first normal stress difference $N_1(\dot\gamma)$ and the dynamic data has also been attempted (see, for example, Laun 1986, Al-Hadithi et al. 1988). Figure 6.1 for a concentrated polymer solution contains the relevant correlation based on the empirical formula of Al-Hadithi et al. (1988), viz.

$$|G_c| = G'\left(1 + \frac{(\eta_0 + \eta')G'}{2\omega(\eta')^2}\right)^{1/2}. \tag{6.3}$$

These authors proposed (as the Cox–Merz equivalent) that $N_1/2$ should be the same function of $\dot\gamma$ as $|G_c|$ is of frequency $\omega$.

Another useful empiricism is to plot the first normal stress difference $N_1$ as a function of shear stress $\sigma$ on logarithmic scales. Often, a master curve is obtained which is independent of temperature. The data usually lie on a reasonably straight line even in the shear-thinning region. Figure 6.12 contains an example of this procedure for a polymer-thickened oil (see also Fig. 4.4).

## 6.11 Practical applications

### 6.11.1 Polymer processing

Nowadays there is no need to emphasize the importance of plastics to modern life: it is regarded as self evident. Polymers are now used as primary materials of construction and have become indispensible.

Specialized texts on rheological aspects of polymer processing have been written by Han (1976), Middleman (1977), Cogswell (1981) and Pearson (1985). Further,

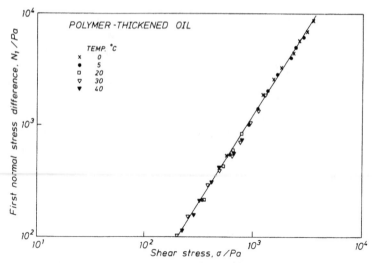

Fig. 6.12 Plot of $N_1$ against $\sigma$ for a polymer-thickened oil. The data for different temperatures fall approximately on the same straight line. The slope is 1.59.

many standard rheological texts have important chapters on the subject (see, for example, Tanner 1985).

In the main, processing is carried out on the melt, although in some cases, such as in the formation of films and fibre of heat-sensitive polymers, a solution is processed. The extruder is the most important melt processing machine and it incorporates a number of sub-processes, such as melting and mixing, prior to that of forming into rods, tubes or films. Extruders are also attached to moulds if the required final shape is more complicated. The process is then either injection moulding for "solid" objects or blow moulding when an extrudate sheet is forced by air pressure against a cooled mould surface.

Extruded films and fibres are stretched to achieve their final thickness and molecular orientation. The process of drawing-down film extruded in the form of a tube or bubble is called film blowing, whilst the drawing-down of fibre is called spinning.

Pearson (1985) points out that in many processes the molten polymer can be considered to be an inelastic non-Newtonian liquid and the power-law model is often adequate in process modelling. This view is justified by the fact that many processing flows are characterized by a narrow dimension normal to the direction of flow and the so-called "lubrication approximation" can be employed (see also Tanner 1985). Under these conditions, it is not difficult to argue that shear viscosity is the dominant rheological influence.

However, there are some important processing stiuations, e.g. film blowing, spinning and flow through a contraction, where *extensional* viscosity is clearly the rheological function of importance (cf. Chapter 5). Not surprisingly, therefore, the measurement of the extensional viscosity of polymeric liquids is viewed as a study of industrial importance.

Melt flow instabilities such as "shark-skin" and 'melt fracture' can occur in extrusion and these effectively place upper bounds on operating conditions. A review of such instabilities is provided by Petrie and Denn (1976). There is recent evidence that the materials used to construct processing equipment can have a significant effect on the critical conditions for the onset of melt flow instabilities thus indicating that interfacial as well as viscoelastic effects are important (Rama-murthy 1986).

### 6.11.2 Polymers in engine lubricants

For more than 40 years, polymers have been used to make the so-called multigrade oils. The motivation is to reduce the large variation of viscosity with temperature and to thereby maintain good hydrodynamic lubrication at high temperatures, without incurring excessive frictional losses at low temperatures. The viscosity level of an automotive oil is specified by a grade number. By the use of polymer, it has become possible to formulate oils which meet the requirements of more than one grade, something which is impossible with ordinary lubricating-oil fractions. Hence the term "multigrade". The polymer additives are known as Viscosity-Index (or VI) improvers.

The thickening effect of the VI improvers depends on the grades required, but in modern oils it can amount to more than a tripling of the base-oil viscosity. The blends show shear-thinning behaviour, which does not become significant until shear rates exceed about $10^5$ s$^{-1}$. Since engine bearings can operate at shear rates two orders of magnitude higher than this, shear-thinning must be accommodated in the design of a lubricant.

Normal stresses and other viscoelastic effects can be measured in multigrade oils and evidence is accumulating that viscoelasticity, together with the increased viscos-ity, may help in carrying the bearing load. These and many other aspects of the rheology of lubrication were reviewed by Hutton (1980) although, at that time, viscoelasticity was thought to have a relatively minor role to play in journal bearings.

### 6.11.3 Enhanced oil recovery

Unlike a water reservoir, a crude oil reservoir is located in the pores of sedimentary rock and the extraction of oil is not easy. A newly tapped deposit is held under earth pressure which will drive a proportion of the oil to the receiving bore holes. When this begins to fail, the secondary recovery process is initiated. This involves pumping water down holes set in a ring around the field. The water sweeps crude before it, but, since the water has the lower viscosity, the water/oil interface is unstable. When the instability occurs, water bypasses the oil in the form of "fingers" and reaches the receiving well first: any variation in the porosity of the oil-bearing strata can exacerbate the fingering effect. Oil wells which have been abandoned because of fingering may still contain as much as 50% of the original deposit. Hence, tertiary or enhanced oil recovery (EOR) is potentially important.

Many proposed EOR methods are being researched. One method is known as polymer flooding. The principle is to stabilize the water/oil interface by using an aqueous polymer solution as the displacement fluid. Candidates for use in this connection are the relatively rigid Xanthan gum and the more flexible polyacrylamide. We have already shown in §5.5 that aqueous solutions of these polymers can behave similarly in shear flow but in a qualitatively different fashion in extensional flow. This may be important in the relatively severe conditions near the well bore (see, for example, Walters and Jones 1988).

### 6.11.4 Polymers as thickeners of water-based products

Many consumer products require the use of a polymer thickener. This might be for some technical reason such as the suspension of fine particles: typical are kaolin in medicines and abrasive particles in liquid abrasive cleaners. Although the amount of thickening is of paramount importance, the elasticity and the extensional viscosity are also relevant. The visual presence of "wobbliness" (which indicates high elasticity) and "stringiness" (which indicates a high extensional viscosity) are sometimes unacceptable to the consumer. It is of interest to note that polyacrylamide, one of the polymers of Figs. 5.18 and 5.19, is unacceptable in this respect because it forms solutions which have high elasticity and a high extensional viscosity. On the other hand the other polymer, Xanthan gum, forms solutions which are not so highly elastic and have a much lower extensional viscosity. Consequently, on the basis of equal shear viscosity, Xanthan gum solutions are not wobbly or stringy in comparison with polyacrylamide solutions. Toothpastes and paints require the use of such "inelastic" polymer thickeners, whilst the food industry relies heavily on such materials for thickening sauces and soups. Polysaccharide-type polymers and natural gums are often used for this purpose (Glicksman 1969).

CHAPTER 7

# RHEOLOGY OF SUSPENSIONS

## 7.1 Introduction

The rheology of suspensions has been the subject of serious research for many years, mainly because of its obvious importance in a wide range of industrial applications (see, for example, Barnes 1981). Suspensions include cement, paint, printing inks, coal slurries, drilling muds and many proprietory products like medicines, liquid abrasive cleaners and foodstuffs. Examples of suspensions where the particles are deformable range from emulsions to blood.

### 7.1.1 The general form of the viscosity curve for suspensions

The general viscosity/shear rate curve for all suspensions is shown schematically in Fig. 7.1. We could anticipate most of this behaviour from the general discussion of Chapter 2. The first Newtonian plateau at low shear rate is followed by the power-law shear-thinning region and then by a flattening-out to the upper (second) Newtonian plateau. At some point, usually in this upper Newtonian region, there can be an increase in viscosity for suspensions of solid particles, given the appropriate conditions. In certain situations the first Newtonian plateau is sometimes so high as to be inaccessible to measurement. In such cases the low-shear rate behaviour is often described by an apparent yield stress.

The factors controlling the details of the general flow curve in particular cases will now be considered. One point worth stressing here is that the relevant measure of the amount of material suspended in the liquid is that fraction of space of the

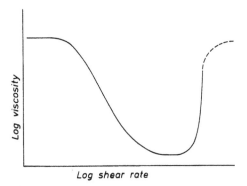

Fig. 7.1 Schematic representation of the flow curve of a concentrated suspension.

total suspension that is occupied by the suspended material. We call this the phase volume $\phi$. This is the volume-per-volume fraction, and not the weight-per-weight fraction that is often used in defining concentration. The reason why phase *volume* is so important is that the rheology depends to a great extent on the hydrodynamic forces which act on the surface of particles or aggregates of particles, generally irrespective of the particle density.

### 7.1.2 Summary of the forces acting on particles suspended in a liquid

Three kinds of forces coexist to various degrees in flowing suspensions. First, there are those of colloidal origin that arise from interactions between the particles. These are controlled by properties of the fluid such as polarisability, but not by viscosity. These forces can result in an overall *repulsion* or *attraction* between the particles. The former can arise, for instance, from like electrostatic charges or from entropic repulsion of polymeric or surfactant material present on the particle surfaces. The latter can arise from the ever-present London–van der Waals attraction between the particles, or from electrostatic attraction between unlike charges on different parts of the particle (e.g. edge/face attraction between clay particles). If the net result of all the forces is an attraction, the particles tend to flocculate, whilst overall repulsion means that they remain separate (i.e. dispersed or deflocculated).

Each colloidal force has a different rate of decrease from the surface of the particle and the estimation of the overall result of the combination of a number of these forces operating together can be quite complicated. Figure 7.2 shows the form of some single and combined forces (see Hunter 1987 for a more detailed account of colloidal forces).

Secondly, we must consider the ever-present Brownian (thermal) randomising force. For particles of all shapes, this constant randomisation influences the form of the radial distribution function (i.e. the spatial arrangement of particles as seen from the centre of any one particle), whereas for non-spherical particles, spatial orientation is also being randomised. The Brownian force is of course strongly size-dependent, so that below a particle size of 1 $\mu$m it has a big influence. This force ensures that the particles are in constant movement and any description of the spatial distribution of the particles is a time average.

Thirdly, we must take into account the viscous forces acting on the particles. The viscous forces are proportional to the local velocity difference between the particle and the surrounding fluid. Hence the way these affect the suspension viscosity enters via the viscosity of the continuous phase which then scales all such interactions. For this reason, "suspension viscosity" is usually considered as the viscosity relative to that of the continuous phase.

Clearly, the rheology measured macroscopically is strongly dependent on these microstructural considerations. For instance, the presence of isolated particles means deviation of the fluid flow lines and hence an increased viscosity. At higher concentrations, more resistance arises because particles have to move out of each other's way. When particles form flocculated structures, even more resistance is encountered because the flocs, by enclosing and thus immobilising some of the

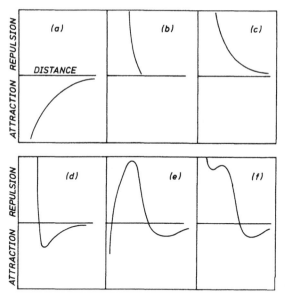

Fig. 7.2 Examples of the typical interaction forces between a pair of sub-micron particles: (a) van der Waals attraction (omnipresent); (b) Steric repulsion due to adsorbed macromolecules; (c) Electrostatic repulsion due to the presence of like charges on the particles and a dielectric medium; (d) A combination of (a) and (b); (e) A combination of (a) and (c); (f) A combination of (a), (b) and (c).

continuous phase, have the effect of increasing the apparent phase volume, thus again giving a higher than expected viscosity.

### 7.1.3 Rest structures

When particles are introduced into a liquid at rest they usually assume a state of thermodynamic equilibrium which, when the Brownian motion dominates, corresponds to a random disordered state.

When colloidal forces dominate, the particles form structures whose forms depend on whether the overall forces are attractive or repulsive. When they are attractive they form aggregates and when they are repulsive they form a pseudo-lattice.

The particular shape of aggregates can vary from near-spherical flocs to strings. The latter is sometimes referred to as a string-of-pearls structure. Pigment dispersions form flocs and silica dispersions can form the string structures.

Pseudo-lattices are formed by overall repulsion, for instance in systems of particles carrying electrostatic charges of the same sign dispersed in a polar continuous phase. The particles then take up positions as far from each other as possible. If the charge on such particles is very large, movement of the particles is severely restricted and the structure can be visualised as a pseudo-crystal: if the lattice spacing is comparable to optical wavelengths, interference effects occur and the suspension displays irridescence.

Any of the structures mentioned above can be modified by the adsorbtion of surfactant materials onto the surface of the particles. The structures formed by electrostatic charges can be modified by the addition of electrolytes.

The structures discussed so far are formed by near-spherical particles. However, if the basic particle is itself anisotropic, very complicated structures can be formed. One example is an aqueous suspension of bentonite clay: the basic particle is plate-like and it carries charges of opposite sign on the faces and edges. The aggregated structure is then like a house of cards with edges attracted to faces. Another example is a suspension of soap crystals. Soap can be made to crystallize in the form of long ribbons, which then intertwine to form an entangled structure, as in lubricating grease.

A useful method for judging the importance of colloidal forces has been derived by Woodcock (1985). It gives the average distance $\bar{h}$ between first neighbours in terms of the particle size $d$ and the volume fraction $\phi$ as follows:

$$\frac{\bar{h}}{d} = \left[ \left( \frac{1}{3\pi\phi} + \frac{5}{6} \right)^{\frac{1}{2}} - 1 \right]. \tag{7.1}$$

This expression is plotted in Fig. 7.3 for four sizes of particle. The diagram also shows the range of action of typical colloidal forces. This diagram indicates for example that, for a suspension with a $\phi$ of 0.2 and particle size 0.05 $\mu$m, electrostatic interactions will be very important. This will not be so if particles are larger or if the concentration is lower.

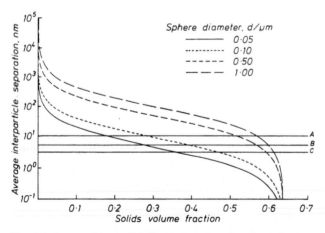

Fig. 7.3 Average interparticle separation as a function of concentration for monodisperse spheres (according to eqn. (7.1)) plotted for four particle sizes. The horizontal lines show twice the distance over which various interparticle forces typically operate: (A) Electrostatic forces in aqueous suspensions with low salt levels; (B) Steric forces originating from adsorbed macromolecules; (C) Steric forces originating from adsorbed nonionic detergents.

### 7.1.4 Flow-induced structures

We shall consider first, relatively unaggregated systems in which the Brownian forces dominate. When a concentrated suspension of this type flows at very low shear rates, the particles necessarily have to move around each other or "bounce off" each other for overall flow to occur. This involves a large resistance and the resulting viscosity is high. On the other hand, the distribution of particles remains essentially undisturbed because the effect of Brownian motion dominates the shear motion and restores the randomness of the rest-state distribution. The viscosity remains essentially constant. At slightly higher shear rates, the imposed velocity gradient induces an orientation of the particle structure, which is not restored by the Brownian motion. However this orientation enables particles to move past each other more freely than at very low shear rates and hence the viscosity is lower. At even higher shear rates, the structure is so grossly orientated that the particles form layers separated by clear layers of the continuous phase. The viscosity is then at its minimum value. The suspension is shear-thinning. The existence of particle layers has been confirmed by light diffraction.

When shearing is stopped the flow-induced layered structure gradually disappears. If the shear stress is increased above a critical value, the layers disrupt and gradually disappear. Hence, the viscosity begins to rise again and it also increases with time of shearing.

Flow-induced structures can also be formed by the more complicated clay and soap suspensions mentioned earlier. In these cases, flow causes the plates and ribbons to align in the direction of flow. This orientation can be detected by optical techniques.

Examples of increased flocculation caused by flow are also known (see, for example, Cheng 1973).

## 7.2 The viscosity of suspensions of solid particles in Newtonian liquids

### 7.2.1 Dilute dispersed suspensions

A considerable amount of progress has been made in predicting the viscosity of dilute suspensions (10% and less phase volume). All studies essentially extend the work of Einstein (1906, 1911) on spheres, so that particle shape, charge and the small amount of hydrodynamic interaction arising when any one particle comes into the vicinity of another can all be taken into account.

Einstein showed that single particles increased the viscosity of a liquid as a simple function of their phase volume, according to the formula

$$\eta = \eta_s(1 + 2.5\phi), \tag{7.2}$$

where $\eta$ is the viscosity of the suspension and $\eta_s$ is the viscosity of the suspending medium.

We notice immediately that in eqn. (7.2) there is no effect of particle size, nor of particle position, because the theory neglects the effects of other particles. When interactions between particles are included, the situation becomes more complicated. The presence of other particles is accounted for by higher-order terms in $\phi$. However the only tractable theory is for extensional flow, because only in this type of flow can the relative position of the particles be accounted for analytically. Batchelor (1977) gives the viscosity in this case as

$$\eta = \eta_s \left(1 + 2.5\phi + 6.2\phi^2\right),\tag{7.3}$$

where the viscosities must now to be interpreted as extensional viscosities (Chapter 5).

A number of experimental determinations of the term multiplying $\phi^2$ for *shear flows* have been made, but the range of values so obtained is large (varying from about 5 to 15).

A great deal of work has been done and many reviews written (see, for example, Barnes 1981) on dilute suspensions, but almost all conclude that, apart from providing some limiting condition for the concentrated case, the work is of little relevance to suspensions of industrial importance. Dilute suspension theory covers the range below 10% phase volume, and this accounts for no more than a 40% increase in viscosity over the continuous phase.

### 7.2.2 Maximum packing fraction

The influence of particle concentration on the viscosity of the concentrated suspensions is best determined in relation to the maximum packing fraction. There must come a time, as more and more particles are added, when suspensions "jam up", giving continuous three-dimensional contact throughout the suspension, thus making flow impossible, i.e. the viscosity tends to infinity. The particular phase volume at which this happens is called the maximum packing fraction $\phi_m$, and its value will depend on the arrangement of the particles. Examples are given in Table 7.1. Maximum packing fractions thus range from approximately 0.5 to 0.75 even for monodisperse spheres.

The maximum packing fraction, as well as being controlled by the type of packing, is very sensitive to particle-size distribution and particle shape (see, for

TABLE 7.1
The maximum packing fraction of various arrangements of monodisperse spheres

| Arrangement | Maximum packing fraction |
| --- | --- |
| Simple cubic | 0.52 |
| Minimum thermodynamically stable configuration | 0.548 |
| Hexagonally packed sheets just touching | 0.605 |
| Random close packing | 0.637 |
| Body-centred cubic packing | 0.68 |
| Face-centred cubic/ hexagonal close packed | 0.74 |

example, Wakeman 1975). Broader particle-size distributions have higher values of $\phi_m$ because the smaller particles fit into the gaps between the bigger ones. On the other hand, nonspherical particles lead to poorer space-filling and hence lower $\phi_m$. Particle flocculation can also lead to a low maximum packing fraction because, in general, the flocs themselves are not close-packed.

From the above considerations, we see that the ratio $\phi/\phi_m$ is a relevant normalized concentration.

### 7.2.3 Concentrated Newtonian suspensions

The situation for concentrated suspensions, where we expect higher-order terms than $\phi^2$ to be important, is even more difficult to analyse from a theoretical point of view. The only methods available to tackle the problem are to introduce a technique for averaging the influence of neighbouring particles or alternatively to simulate the situation using computer modelling.

One recent development, based on an averaging technique, is that of Ball and Richmond (1980) who essentially start from the assumption that the effect of all the particles in a concentrated suspension is the sum of the effects of particles added sequentially. Hence the Einstein equation can be written in a differential form

$$d\eta = (5\eta/2)\, d\phi, \tag{7.4}$$

where $d\eta$ is the increment of viscosity on the addition of a small increment of phase volume $d\phi$ to a suspension of viscosity $\eta$. The viscosity of the final suspension is then obtained by integrating the phase volume between 0 and $\phi$, for which the viscosity is $\eta_s$ and $\eta$, respectively. Then

$$\eta = \eta_s \exp(5\phi/2). \tag{7.5}$$

Ball and Richmond point out that this omits the correlations between spheres due to their finite size. This means that when a particle is added to a relatively concentrated suspension it requires more space than its volume $d\phi$, due to packing difficulties. Therefore $d\phi$ has to be replaced by $d\phi/(1 - K\phi)$, where $K$ accounts for the so-called "crowding" effect. Integration now yields

$$\eta = \eta_s (1 - K\phi)^{-5/(2K)}. \tag{7.6}$$

From this equation we see that the viscosity becomes infinite when $\phi = 1/K$. Therefore, we can identify $1/K$ with the maximum packing fraction $\phi_m$. Ball and Richmond's expression is effectively identical to that of Krieger and Dougherty (1959). Krieger and Dougherty's theory also states that, in the general case, the $5/2$ factor could be replaced by the intrinsic viscosity $[\eta]$. * The value of $5/2$ is the

---

* Note: in suspension rheology $[\eta]$ is dimensionless, since the phase volume is also dimensionless (see eqn. (7.2)); whereas in polymer rheology the concentration is usually expressed as a mass per unit volume, thereby giving $[\eta]$ the dimensions of a reciprocal concentration (see § 6.5).

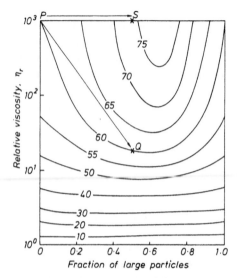

Fig. 7.4 Effect of binary particle-size fraction on suspension viscosity, with total % phase volume as parameter. The particle-size ratio is 5 : 1. P → Q illustrates the fiftyfold reduction in viscosity when a 60% v/v suspension is changed from a mono- to a bimodal (50/50) mixture. P → S illustrates the 15% increase in phase volume possible for the same viscosity when a suspension is changed from mono- to bimodal.

intrinsic viscosity for an ideal dilute suspension of spherical particles. Replacing it by $[\eta]$ allows particles of any shape to be accounted for.

The Krieger–Dougherty equation is

$$\eta = \eta_s \left(1 - \phi/\phi_m\right)^{-[\eta]\phi_m}. \tag{7.7}$$

Equations 7.6 and 7.7 both reduce to the Einstein equation (eqn. (7.2)) when $\phi$ is small.

The values of $\phi_m$ obtained from the empirical use of eqn. (7.7) are strongly dependent on the particle-size distribution. Thus, $\phi_m$ increases with increasing polydispersity (i.e. the spread of sizes). This is illustrated by Fig. 7.4 where the viscosities of mixtures of large and small particles are plotted as a function of the total phase volume. The large reduction in viscosity seen near a fraction of 0.6 of large particles is known as the Farris effect. The effect is very large at a total phase volume of more than 50%. Mixing particle sizes thus allows the viscosity to be reduced whilst maintaining the same phase volume, or alternatively, the phase volume to be increased whilst maintaining the same viscosity. Similar effects can also be shown for tertiary mixtures (cf. Fig. 7.5). In the example shown in Fig. 7.5 the minimum relative viscosity is approximately 25 for the optimum tertiary mixture and is over 30 for the binary mixture. All these effects can be predicted using eqn. (7.7) by assuming, for instance, that the small particles thicken the continuous phase

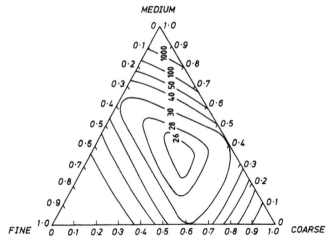

Fig. 7.5 Effect of particle-size distribution on trimodal suspension viscosity. Contours show values of the relative viscosity at 65% total solids (from the theoretical relationship of Farris (1968).

and the next-size-up particles then thicken this phase; the result for a binary mixture being

$$\eta = \eta_s(1 - \phi_1/\phi_{m1})^{-[\eta_1]\phi_{m1}}(1 - \phi_2/\phi_{m2})^{-[\eta_2]\phi_{m2}}. \tag{7.8}$$

Most suspensions of industrial interest have a continuous distribution of particle sizes which often fit some empirical mathematical expression. However no information is available in the rheological literature on how the parameters of such a fit control $\phi_m$. Each system has therefore to be measured and $\phi_m$ found by nonlinear curve fitting. Once $\phi_m$ is found for any practical suspension, it is a useful parameter to assess the effect on viscosity of changing the dispersed phase concentration or the continuous phase viscosity.

Thus far, we have concentrated on the effect of spherical particles on the viscosity of suspensions. However, particle asymmetry has a strong effect on the intrinsic viscosity and maximum packing fraction, and hence on the concentration/viscosity relationship. A number of studies have shown that any deviation from spherical particles means an increase in viscosity for the same phase volume. Figures 7.6 and 7.7 illustrate this point. It will also be seen that, generally speaking, rods have a greater effect than discs in increasing the viscosity. This is in accordance with theory, at least as far as it goes for dilute suspensions. Barnes (1981) provides simple empiricisms for the effect of very large axial ratio on the intrinsic viscosity $[\eta]$. These are

discs: $[\eta] = 3$ (axial ratio)$/10$, \hfill (7.8a)

rods: $[\eta] = 7\left[\text{(axial ratio)}^{5/3}\right]/100$. \hfill (7.8b)

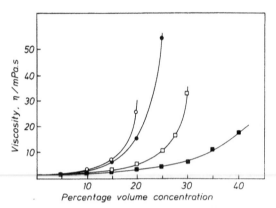

Fig. 7.6 Dependence of the viscosity of differently shaped particles in water on concentration at a shear rate of $300s^{-1}$ (from Clarke 1967). (■) spheres; (□) grains; (●) plates; (○) rods. (See Table 7.2.)

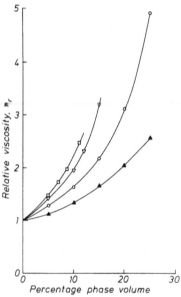

Fig. 7.7 Dependence of the relative viscosity of glass fibre suspensions of various length/diameter ratios ($L/D$) (cf. Giesekus 1983). (▲▲▲) spheres; (○○○) $L/D = 7$; (▽▽▽) $L/D = 14$; (□□□) $L/D = 21$ (see Table 7.2).

Table 7.2 gives the values of $[\eta]$ and $\phi_m$ obtained by fitting the results of a number of experimental investigations on suspensions of asymmetric particles using eqn. (7.7). The trend to higher $[\eta]$ and lower $\phi_m$ with increasing asymmetry is clearly seen, but the product of the two terms changes little. This fact has practical value in making estimates of the viscosity of a wide variety of suspensions. The values of $[\eta]$ are qualitatively in line with the predictions of eqns. (7.8a) and (7.8b).

TABLE 7.2
The values of $[\eta]$ and $\phi_m$ for a number of suspensions of asymmetric particles, obtained by fitting experimental data to eqn. (7.7)

| System | $[\eta]$ | $\phi_m$ | $[\eta]\phi_m$ | Reference |
|---|---|---|---|---|
| Spheres (submicron) | 2.7 | 0.71 | 1.92 | de Kruif et al. (1985) |
| Spheres (40 $\mu$m) | 3.28 | 0.61 | 2.00 | Giesekus (1983) |
| Ground gypsum | 3.25 | 0.69 | 2.24 | Turian and Yuan (1977) |
| Titanium dioxide | 5.0 | 0.55 | 2.77 | Turian and Yuan (1977) |
| Laterite | 9.0 | 0.35 | 3.15 | Turian and Yuan (1977) |
| Glass rods | 9.25 | 0.268 | 2.48 | Clarke (1967) |
| ($30 \times 700$ $\mu$m) | | | | |
| Glass plates | 9.87 | 0.382 | 3.77 | Clarke (1967) |
| ($100 \times 400$ $\mu$m) | | | | |
| Quartz grains | 5.8 | 0.371 | 2.15 | Clarke (1967) |
| ($53-76$ $\mu$m) | | | | |
| Glass fibres: | | | | |
| axial ratio-7 | 3.8 | 0.374 | 1.42 | Giesekus (1983) |
| axial ratio-14 | 5.03 | 0.26 | 1.31 | Giesekus (1983) |
| axial ratio-21 | 6.0 | 0.233 | 1.40 | Giesekus (1983) |

### 7.2.4 Concentrated shear-thinning suspensions

Although the theory described above (see eqn. (7.6)) was derived for the spherically symmetrical radial distribution function, i.e. the very low shear rate case, it has been found to work surprisingly well over a range of shear rates for which the structure is anistropic. It accommodates these situations by allowing $[\eta]$ and $\phi_m$ to vary with shear rate, thus accounting for shear-thinning by the fact that the flow brings about a more favourable arrangement of particles. The tendency to form two-dimensional structures rather than three is one such favourable rearrangement.

Considering first the viscosity/phase volume relationships at very low and very high shear rates, it is found that they both fit the Krieger–Dougherty equation.

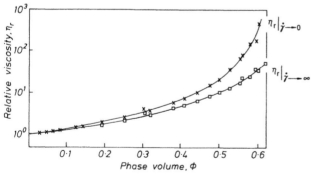

Fig. 7.8 Relative viscosity versus phase volume for monodisperse latices. Data points are those of Krieger (1972) and de Kruif et al. (1985) combined. The upper line relates to the zero shear-rate asymptotic relative viscosity, and is the best fit to the Krieger–Dougherty eq. (7.7) with $\phi_m = 0.632$ and $[\eta] = 3.133$. The lower line relates to the high shear-rate asymptotic value of 'relative viscosity' and is the best fit to eqn. (7.7), with $\phi_m = 0.708$ and $[\eta] = 2.710$.

Combining data from Krieger (1972) and de Kruif et al. (1985) (Fig. 7.8) the values of $\phi_m$ and $[\eta]$ pertinent to the two situations are

$$\phi_m(\dot{\gamma} \to 0) \quad = 0.632, \quad [\eta](\dot{\gamma} \to 0) = 3.13,$$

$$\phi_m(\dot{\gamma} \to \infty) = 0.708, \quad [\eta](\dot{\gamma} \to \infty) = 2.71.$$

Secondly, a number of workers have found that not only can viscosity be related to phase volume at the extremes of shear rate, but at intermediate values as well. However, it is usually necessary in this case to correlate values of viscosity measured at the same shear stress, not shear rate.

A further reduction of data is possible following the suggestion of Krieger (1972). He recognised that although $\phi_m$ and $[\eta]$ are stress-dependent, they are independent of particle size. In order to relate the viscosity of different particle-size suspensions, he suggested that, instead of shear rate, a new variable be used, namely

$$P_e = 6\sigma a^3 / kT, \tag{7.9}$$

where $a$ is the particle radius, $\sigma$ the shear stress and $kT$ the usual unit of thermal energy.

Krieger was able to show that for noninteracting suspensions, the viscosity/shear stress curves for sub-micron suspensions are reducible to a single curve (Fig. 7.9) whatever the particle size, temperature and continuous phase viscosity. The "Krieger variable" is dimensionless and is in fact a modified Péclet number *. Krieger realised that the viscosity of the suspension is more relevant than that of the continuous phase in accounting for concentrated suspensions. The shape of the curve of viscosity versus $P_e$ in most cases follows the empirical Ellis model:

$$\frac{\eta - \eta_\infty}{\eta_0 - \eta_\infty} = \frac{1}{1 + b(P_e)^p}, \tag{7.10}$$

where $b$ and $p$ are dimensionless quantities (see §2.3.2 and the footnote on p. 18).

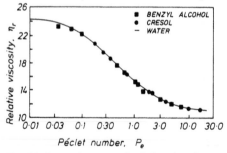

Fig. 7.9 Composite curve of relative viscosity versus modified Péclet number (see eqn. (7.9)). Points are for 0.1–0.5 $\mu$m latex particles dispersed in two solvents; the solid line is for the same-sized particles dispersed in water; the volume fraction is 0.50 (reproduced from Krieger 1972).

---

* The Péclet number is the ratio of the viscous force experienced by a particle to the Brownian force.

### 7.2.5 Practical consequences of the effect of phase volume

#### I Weighing errors

An examination of eqn. (7.7) will show that, at phase volumes less than about 30% concentration, the viscosity changes slowly with concentration, and the viscosities at very low and very high shear rates, respectively, are essentially the same, i.e. dilute suspensions are basically Newtonian. However, at values of $\phi$ around 0.5, small changes in either $\phi$ or $\phi_m$ give large changes in viscosity and alter the degree of shear-thinning. Thus, small errors in the weighing involved in the incorporation of the particles making up the suspension can give very large variations in viscosity, (see, for example, Fig. 7.10). Also, as is often the case in polymer latices after their manufacture, the absorption of small amounts of the suspending phase into the suspended phase can cause large changes in the viscosity of concentrated suspensions. Even half of one percent lost from the continuous phase to the suspended phase means an increase in the phase volume of one percent, which can, in the case of high phase volume, lead to a doubling or more of the viscosity! When it is realised that this is equivalent to the particle size increasing by about 0.2%, and that particle size measurement techniques are nowhere near able to detect such small changes, we see that viscosity is a very sensitive variable in suspension rheology.

#### II Effect of phase separation

Any sedimentation or 'creaming' of the particles in a viscometer will result in an increase in the indicated viscosity. At phase volumes above about 0.5 this effect is extremely pronounced.

#### III Wall effect

Yet another factor which must be taken into account in rheometry, as well as in industrial situations, is known as the "wall effect". This term covers the phenomena

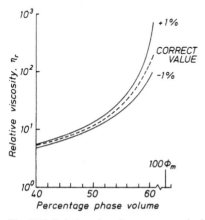

Fig. 7.10 Relative viscosity versus nominal phase volume showing the effect of a 1% error in phase volume.

which lead to a reduction in concentration of a suspension adjacent to the solid wall of the flow channel. One such phenomenon arises from the geometric impossibility of arranging particles near a wall in the same way as they are arranged in the bulk. Another is a particle migration from regions of high shear rate to regions of low shear rate. The hydrodynamic redistribution of particles demonstrated by Segré and Silberberg (1962) is yet another.

A number of approaches have been attempted to account for the wall depletion. Most use the concept of a layer of continuous phase only at the wall and the normal dispersion everywhere else, with the thickness of the depleted layer being of the order of the radius of the particles.

The result of the wall effect in tube viscometers is a reduction in the measured viscosity, the reduction increasing as the tube radius is decreased. Figure 7.11 illustrates these effects for lubricating greases.

### 7.2.6 Shear-thickening of concentrated suspensions

Given the correct conditions, all concentrated suspensions of non-aggregating solid particles will show shear-thickening. The particular circumstances and severity of shear-thickening will depend on the phase volume, the particle-size distribution and the continuous phase viscosity. The region of shear-thickening usually follows that of the shear-thinning brought about by two-dimensional layering. The layered arrangement is unstable, and is disrupted above a critical shear stress. The ensuing random arrangement increases the viscosity. The effect has been studied using an optical diffraction system (Hoffman 1972). The result in terms of viscosity/shear rate for a range of particle concentrations is shown in Fig. 7.12.

A number of studies have shown that the critical shear rate for transition to shear-thickening varies little with phase volume when the phase volume is near 0.50. However, at phase volumes much higher than this, the critical shear rate decreases, whilst at phase volumes significantly below 0.5, the opposite is true (see Fig. 7.13).

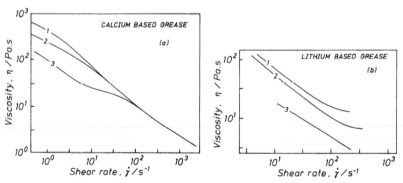

Fig. 7.11 (a) and (b): Viscosity measurements for greases measured in axial flow between parallel cylinders (plunger viscometer) using various annular gaps. Gap/mm: (*1*) 0.624; (*2*) 0.199; (*3*) 0.042. The viscometer-size effect disappears at higher shear rates away from the "yield stress" region. (Bramhall & Hutton 1960).

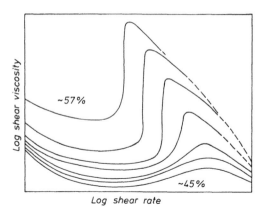

Fig. 7.12 Schematic representation of viscosity versus shear rate for shear-thickening systems, with phase volume as parameter (cf. Barnes, 1989).

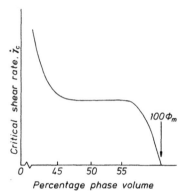

Fig. 7.13 Schematic representation of the dependence of the critical shear rate for the onset of shear-thickening $\dot{\gamma}_c$, as a function of the phase volume of the dispersed phase $\phi$.

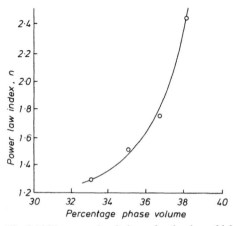

Fig. 7.14 The power-law index $n$ for the shear-thickening region of starch suspensions (after Griskey and Green 1971).

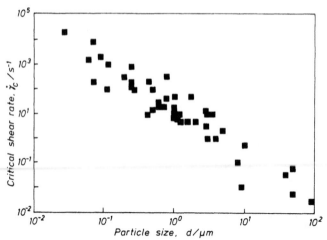

Fig. 7.15 Shear rate for the onset of shear-thickening $\dot{\gamma}_c$, versus particle size for suspensions with phase volumes around 0.5 (Barnes, 1989).

The level and slope of the viscosity/shear rate curve above the transition increases with increase in particle concentration (see Figs. 7.12 and 7.14). Evidence is accumulating that the viscosity decreases again at very high shear rates, although experiments are difficult to carry out at very high phase volume due to flow instability. The ultimate decrease is readily observed at lower phase volume.

For any phase volume around 0.50, it is found that the effect of particle size on the value of the critical shear rate is quite large. In fact, it is approximately proportional to the inverse of the square of the particle size (see Fig. 7.15). The viscosity of the continuous phase is also very important, and an increase in this viscosity decreases the critical shear rate. This reflects the greater relevance of shear stress (rather than shear rate) for the onset of shear-thickening, (see Fig. 7.16).

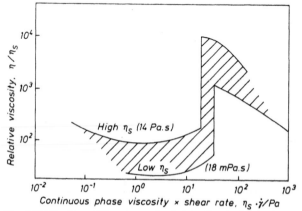

Fig. 7.16 Envelope of flow curves for latices dispersed in various solvents whose viscosities vary from 18 mPa.s to 14 Pa.s. (Redrawn from Hoffman 1972.)

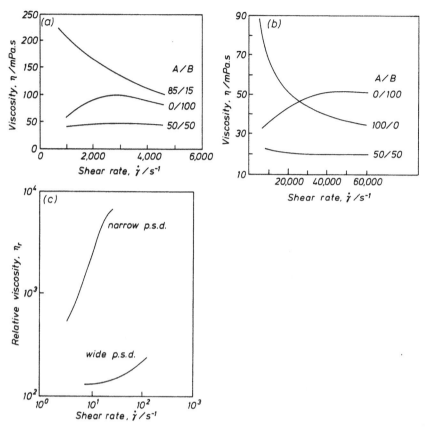

Fig. 7.17 (a) Viscosity of clays A and B at different ratios as a function of shear rate at a solids content of 44% by volume (67% by weight). Clay A 9 $\mu$m, clay B 0.7 $\mu$m. (Alince and Lepoutre 1983); (b) Viscosity of calcium carbonate blends at 48% solids content by volume (71.4% by weight) as a function of shear rate. Clay A 12 $\mu$m, clay B 0.65 $\mu$m. (Alince and Lepoutre 1983); (c) Effect of particle-size distribution. (Redrawn from Williams et al. 1979).

The severity of shear-thickening is often alleviated by widening the particle-size distribution (see Fig. 7.17).

## 7.3 The colloidal contribution to viscosity

### 7.3.1 Overall repulsion between particles

Overall repulsion between the particles of a suspension is created if the particles carry electrostatic charges of the same sign. The particles then take up positions as far from one another as possible. For flow to occur, particles have to be forced out of their equilibrium positions and induced to move against the electric fields of neighbouring particles into nearby vacancies in the imperfect lattice. Goodwin

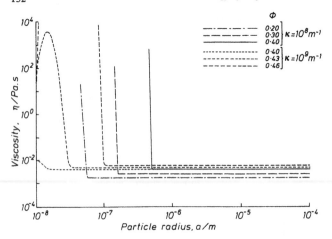

Fig. 7.18 Particle size versus viscosity for a suspension of charged particles, for various particle phase volumes and charge. The viscosity is the sum of the two contributions given by eqns. (7.7) and (7.11). The value of $E^*$ used here is the difference in potential energy between the particle rest-state and the maximum potential it experiences as it "jumps" to the next rest-state site. The particle arrangement is assumed to be face-centred cubic. The charge on the particle is characterised by the value of $\kappa$ which is the inverse of the double-layer thickness. This is a measure of the distance over which the electrostatic potential acts, measured from the particle surface.

(1987) has given the following equation to evaluate the extra contribution of repulsion to the zero shear-rate viscosity over and above the usual Krieger–Dougherty contribution:

$$\Delta \eta = \frac{h}{b^3} \exp(E^*/kT),\tag{7.11}$$

where $h$ is Planck's constant, $b$ is the centre-to-centre particle separation and $E^*$ is the activation energy calculated for self-diffusion.

This expression predicts very large (but finite) viscosities compared to the viscous contribution alone, as accounted for by the Krieger–Dougherty expression, (see Fig. 7.18). The effects of particle size and concentration are very strong, the latter being accounted for in the $E^*$ factor.

Goodwin also suggests that shear-thinning in such systems will become significant at a shear stress of the order of $kT/b^3$.

At very high shear rates, two-dimensional layering occurs and the electrostatic contribution loses its dominance. The viscosity decreases towards that given by the Krieger–Dougherty expression for non-interacting particles.

The range of the electrostatic forces is greatly decreased if electrolyte is added to the solution, so screening the charges on the particles. This means that using electrostatic forces to thicken suspensions is limited to very pure systems, since slight electrolyte contamination can give a large decrease in viscosity.

Repulsion can also arise from the entropic forces caused by the interaction of the chains of any polymer adsorbed onto the particle surfaces. Although not always operating over great distances from the particle, they can affect the viscosity at high phase volumes.

### 7.3.2 Overall attraction between particles

The formation of flocs traps part of the continuous phase, thus leading to a bigger effective phase volume than that of the primary particles. This gives an additional increase in the viscosity over and above that expected from the phase volume of the individual particles. When flocculated suspensions are sheared, the flocs rotate, possibly deform and, if the applied stress is high enough, begin to break down to the primary particles.

Flocs sometimes take the form of chains which form networks throughout the liquid. The length of the strands is a function of the shear stress.

All flocculated structures take time to break down and rebuild, and thixotropic behaviour is usually associated with flocculated suspensions; clay suspensions being classic examples. The driving force to rebuild the floc is Brownian motion, and since this increases with decrease in particle size, the rate of thixotropic change is a function of particle size. Thus, one expects systems of large particles to recover their viscosity slower than systems of small particles. Similarly large-particle suspensions will break down faster under shearing. These considerations are important in the design of thixotropic products such as paints and printing inks.

The attraction between particles can be reduced in a number of ways, including the adsorbtion of molecules onto the surface of the particles. Adding electrolyte to clay suspensions can reduce the differential charge effects. In all these cases, the viscosity is reduced substantially.

Flocculated systems usually have very high viscosities at low shear rate, and are very shear-thinning, often giving the impression of a yield stress (see Fig. 7.19). In many cases the Bingham model has been used to describe their behaviour.

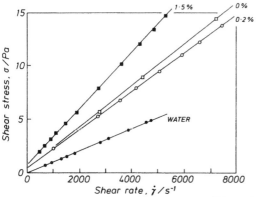

Fig. 7.19 Shear stress/shear rate curves for 1.5% by-weight suspensions of bentonite Supergel in water (0%), and with 0.2% and 1.5% sodium chloride additions. The corresponding curve for pure water is shown for comparison (Ippolito 1980).

## 7.4 Viscoelastic properties of suspensions

Viscoelastic properties of deflocculated suspensions arise from particle interactions of all kinds. If these require a preferred arrangement of the particles at rest in order to fulfil some minimum energy requirement, there will always be a tendency for the suspension to return, or relax, to that arrangement. It is possible to make small perturbations about the preferred state by means of small-amplitude oscillatory experiments. In this case the measure of elasticity is the dynamic rigidity $G'$ (§ 3.5). Figure 7.20 shows a typical example of such behaviour.

Elastic effects are also observable in a steady simple-shear flow through normal

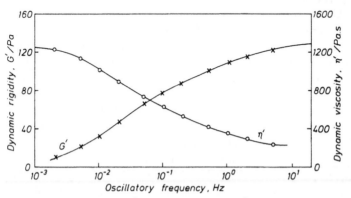

Fig. 7.20 Dynamic viscosity $\eta'$ and dynamic rigidity $G'$ as functions of frequency for polystyrene latex in $10^{-2}$ $M$ NaCl aqueous solution $\phi = 0.35$, particle radius $= 0.037$ $\mu$m. (J. Goodwin, private communication.)

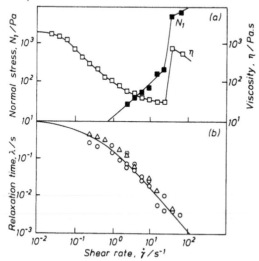

Fig. 7.21 (a) Normal stress and viscosity versus shear rate for a PVC organosol ($\phi = 0.54$) dispersed in dioctyl phthalate; (b) The relaxation time (defined as $\lambda = N_1/\sigma\dot{\gamma}$) plotted against the shear rate derived from the data in (a). (Willey and Macosko 1978).

Elastic effects are also observable in a steady simple-shear flow through normal stress effects (cf. Chapter 4). This is demonstrated for a typical colloidal system in Fig. 7.21. In general, the normal stresses found in colloidal systems are lower than those in comparable polymeric liquids.

## 7.5 Suspensions of deformable particles

Many dispersions are made from deformable particles, the most obvious examples being emulsions and blood. In dealing with the rheology of these systems, all the earlier factors like interparticle forces are relevant, but the effect of phase volume is not so extreme as with solids. The maximum phase volume is usually much higher than with solid particles since the particles deform to accommodate the presence of near neighbours. In this situation, the shape of the particles is polyhedral and the suspension resembles a foam in its structure. Maximum packing fractions of 0.90 and above are usual (Pal et al. 1986).

Figure 7.22 shows the viscosity/shear rate profiles for typical emulsions. The familiar shape seen for solid dispersions is again apparent as is the increasingly non-Newtonian behaviour with increase in concentration. However, the asymptotic value of viscosity at high shear rate is generally much lower than that observed for a dispersion of solid particles at the same phase volume. This effect is ascribable to particle deformation in the emulsion (see, for example, Fig. 7.23 and compare Fig. 7.8).

Whereas many model studies have been carried out on monodisperse solid particle dispersions, the usual mode of production of emulsions by droplet breakup means that making monodisperse droplet-size samples is difficult. Therefore, model studies have not always been able to distinguish between the relative effects of droplet size and the shape of the droplet-size distribution. What is clear however is

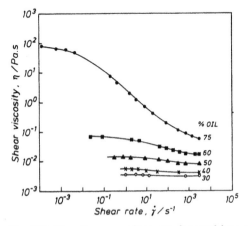

Fig. 7.22 Viscosity versus shear rate for emulsions of silicone oil in water at various values of phase volume of oil.

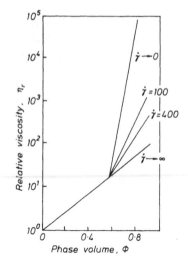

Fig. 7.23 The relative viscosity of a wide-particle-size emulsion (see Pal et al. 1986) for a range of shear rates. Note the onset of shear-thinning is at $\phi \sim 0.55$ for this particular sample; cf. Fig. 7.22 where the emulsion is not so polydisperse and shear-thinning begins at $\phi \sim 0.4$.

that a smaller droplet size and a more monodisperse droplet size both give an increase in viscosity. Since vigorous mixing of emulsions usually gives smaller and more monodisperse particles, increasing the energy input in emulsion manufacture always increases the viscosity.

Theoretical work carried out by Oldroyd (1953) for very dilute emulsions showed that viscoelasticity results from the restoring force due to the interfacial tension between the continuous and disperse phases. The emulsion droplets at rest are spherical, but become ellipsoidal in shear, with a consequent increase in the surface area.

Solid particles stabilised by adsorbed polymers can appear as deformable particles when the particle size is very small (say < 100 nm). In this case the deformable stabilising layer can form a considerable proportion of the real phase volume. The overall effect is that the viscosity is a decreasing function of nominal particle size, when evaluated at constant phase volume (based on uncoated particles). In this case however the phase volume must be adjusted to take account of the stabilising layer. This can be obtained by measuring the viscosity of very dilute suspensions and applying Einstein's equation (eqn. (7.2)).

## 7.6 The interaction of suspended particles with polymer molecules also present in the continuous phase

There are at least four ways in which particles and polymer molecules interact:

*(i)* Neutrally. That is, the polymer merely acts as a thickener for the continuous

phase and the particles as inert fillers. The only overall effect is an increase in viscosity. Examples of this are found in various toothpastes.

*(ii)* For specific polymers (e.g. block copolymers), some part of the polymer molecule adsorbs onto the particle while the other protrudes into the liquid phase. This can have the effect of hindering any flocculation that might take place, especially with very small particles, and hence prevent any ensuing sedimentation or creaming. The interacting polymer chains of adjacent particles overlap and cause entropic repulsion because their local concentration is higher than the average. Emulsion and dispersion stabilisers such as Gum Arabic are examples showing this phenomenon.

*(iii)* Certain polymers have the ability to anchor particles together. This is called "bridging flocculation". They are usually very high molecular-weight macromolecules with groups that attach to the particles by unlike charge attraction. These are deliberately introduced to cause flocculation in separation processes. An example is the use of the polyacrylamide family of polymers in water-clarification plants. The flocs formed by bridging flocculation are relatively strong and can withstand quite high stresses before breaking down.

*(iv)* Some situations arise where polymers in the continuous phase can cause flocculation of the particles. This "depletion flocculation" arises when polymer molecules, because of their finite size, are excluded from the small gap between neighbouring particles. The concentration difference thus caused between the bulk and the gap causes an osmotic pressure difference. This results in solvent leaving the gap, thus pulling the particles together. This in turn means that even more polymer becomes excluded and the effect grows. Eventually the particles are completely flocculated. The floc strength of such a system is relatively small, certainly as compared to *(iii)* above. There must always be a tendency for this to occur in any system, but the time scale of the particle movement and the level of the force makes it possible to ignore it in some circumstances.

## 7.7 Computer simulation studies of suspension rheology

The computer simulations of flowing suspensions have been reviewed by Barnes et al. (1987). The simulations are not dissimilar to computer simulations of simple liquids such as argon and chlorine. Both simulations use the Newtonian equations of motion and a Lennard–Jones type of particle–particle interaction law. The main differences are that the interparticle forces are smaller and the hydrodynamic resistance to motion is much greater for the suspensions.

The techniques of non-equilibrium molecular dynamics (NEMD) consider an assembly of particles which is given an initial set of positions in space and a set of

velocities. Assemblies of between one hundred and a few thousand can be dealt
with. The NEMD calculation is the re-evaluation of the positions and velocities over
a succession of short intervals (see, for example, Heyes 1986). From this information
the pressure (or stress tensor) can be calculated, hence the viscosity and the normal
stress differences.

Given the similarities of approach, it is not surprising that the main results for
simple liquids and suspensions are similar. Following a lower Newtonian region the
systems display shear-thinning. At still higher shear rates the simulations predict
shear-thickening.

Considerable support for the existence of flow-induced structures in the shear-
thinning region is provided by NEMD. The simulations show that suspensions of
spherical particles form two-dimensional layers which break up at the onset of
shear-thickening. The corresponding results in simple liquids are the formation of

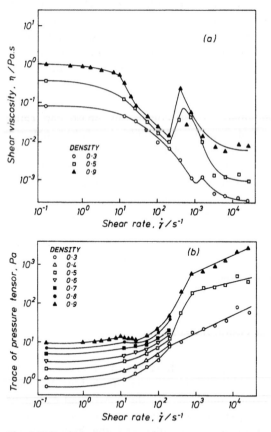

Fig. 7.24 Typical predictions from a computer simulation of a suspension in shear flow (see Barnes,
Edwards and Woodcock 1987). Note: density is normalised using particle parameters. (a) Viscosity versus
shear rate, showing the qualitative features of Fig. 7.1; (b) Shows the trace of the stress tensor. This
osmotic-type pressure results in particle migration to regions of lower shear rate.

strings of molecules aligned in the direction of flow and, for diatomic and longer molecules, an additional alignment of the molecules.

The normal stress components are unequal at shear rates in the shear-thinning region and the differences increase with shear rate. The increase is slow compared with that predicted by polymer theory. Very little normal stress data have been obtained by NEMD and $N_1$ and $N_2$ generally show a considerable scatter. The results are often presented in terms of the trace of the stress tensor. This is also referred to as an osmotic pressure.

Typical results are shown in Fig. 7.24 where the predicted osmotic pressure and viscosity of a dense suspension of submicron particles is shown as a function of shear rate.

The technique of computer simulation is likely to become more important in the future, especially in its ability to study complicated but idealized systems.

# CHAPTER 8

# THEORETICAL RHEOLOGY

## 8.1 Introduction

An alternative title to this chapter could be "Constitutive equations and their uses" since it summarizes the vast majority of published work in theoretical rheology.

The theoretician seeks to express the behaviour of rheologically complex materials through equations relating suitable stress and deformation variables. Such equations are of interest in themselves and *continuum mechanics*, which addresses such matters, is a respectable subject in its own right which occupies the attention of many theoreticians. The relevant equations, called constitutive equations or rheological equations of state, must reflect the materials' microstructure and one fruitful area of study concerns the search for relationships between microstructure and (macroscopic) constitutive equations. We have already touched on this subject in §6.8 and the reader is referred to the important books of Bird et al. (1987(b)) and Doi and Edwards (1986) for further details.

Constitutive equations, which satisfy the basic formulation principles to be discussed in §8.2 and are constrained either by microstructure considerations or by the way the fluids behave in simple (rheometrical) flows, are also used by theoreticians to predict the way the relevant fluids behave in more complex flows of practical importance. Here, the constitutive equations are solved in conjunction with the familiar stress equations of motion and the equation of continuity, subject to appropriate boundary conditions. This is an important branch of non-Newtonian fluid mechanics and is discussed briefly in §8.7.

Practical scientists and engineers may also look to constitutive equations for more modest reasons. For example, they may be interested in reducing experimental data to the knowledge of a small number of material parameters. This can be accomplished by comparing the forms of graphs relating, for example, viscosity/shear rate and/or normal stress/shear rate with the corresponding predictions from likely constitutive equations.

It is not difficult to make out a case for a detailed study of *theoretical rheology* and many books are either devoted to the subject (like that of Truesdell and Noll 1965) or have substantial sections dealing with it (e.g. Lodge 1974, Bird et al. 1987(a), Schowalter 1978, Tanner 1985, Astarita and Marrucci 1974 and Crochet et al. 1984). We have left the subject to the final chapter of the present book for specific reasons, which have already been alluded to in §1.5. The subject is a difficult one by common consent and newcomers to the field (especially those without a mathematical background) are often put off by the complexity of the mathematics. In previous chapters we have attempted to cover the various strands of rheology without involving ourselves with intricate modern continuum mechanics, but it is now time to address this important subject and to point the interested reader in the direction of the many detailed texts and authoritative works on the subject. At the same time, sufficient attention is given to the subject in the present chapter to indicate to the reader, by way of an overview, the most important features of present-day knowledge of continuum mechanics. For most readers this will be all that is required. For others, the growing list of available books on the subject is more than adequate for further detailed study. In this connection we recommend particularly the books by Bird et al. (1987(a)), Schowalter (1978) and Truesdell and Noll (1965). Particular backgrounds and tastes vary greatly and many readers will no doubt obtain significant benefit from the theoretical sections in the other texts mentioned above.

The following discussion is limited to isotropic, time-independent non-Newtonian fluids. Readers interested in anisotropic fluid behaviour should consult the papers of Ericksen and Leslie (see, for example, Leslie 1966, 1979, Jenkins 1978). Those interested in thixotropy and antithixotropy are referred to the review article by Mewis (1979). The amount of *theoretical* literature on time-dependent materials is limited. The considerable conceptual difficulties in the subject are no doubt largely responsible for this deficiency.

## 8.2 Basic principles of continuum mechanics

We seek equations for complex non-Newtonian fluids (with and without fluid memory). To facilitate this, we need to define suitable stress and deformation variables and consistent time-differentiation and integration procedures. The relevant formulation principles required for this purpose are now well-documented and are not controversial. However, a cursory glance at the literature *may* give the reader the mistaken impression that there are two distinct kinds of formulation principles, one associated with the names of Oldroyd, Lodge and their coworkers and the other with the names of Truesdell, Noll, Coleman, Green, Rivlin, Ericksen and their coworkers. Certainly, the way the formulation principles are expressed and applied varies between these groups, but there have been sufficient objective reviews in recent years to prove to the perceptive newcomer to the field that the formulation principles of continuum mechanics have an invariance which is independent of the researcher! They can be expressed in different ways, but there is no essential controversy concerning the two basic approaches referred to and Chapter 2 of the text by Crochet et al. (1984), as well as discussions in other books, should convince the reader of this fact (see also the paper by Lodge and Stark 1972).

We may summarize the formulation principles as follows:

*Principle I.* The constitutive equations must be independent of the frame of reference used to describe them. Expressing the equations in consistent tensorial form will ensure that this principle is automatically satisfied and, in a real sense, this principle is required of all physical theories.

*Principle II.* The constitutive equations must be independent of absolute motion in space (Oldroyd 1950). Any superimposed rigid body motion cannot affect the basic response of the material *.

*Principle III.* The behaviour of a material element depends only on the previous history of that same material element and not on the state of neighbouring elements (Oldroyd 1950). Expressed in an alternative way—the stress is determined by the history of the deformation, and the stress at a given material point is uniquely determined by the history of deformation of an arbitrarily small neighbourhood of that material point (see, for example, Astarita and Marrucci 1974). The basic feature of *principle III* is that "fluid memory" as a concept must be associated with *material elements* and not with *points in space*.

To illustrate how these principles (especially *I* and *III*) apply in situations already discussed, consider again the general integral equations of *linear* viscoelasticity which were introduced in §3.4. These can be written in the tensorial form **

$$
\left.
\begin{aligned}
\sigma_{ik} &= -p\delta_{ik} + T_{ik}, \\
T_{ik}(\underline{x}, t) &= 2\int_{-\infty}^{t} \phi(t - t')d_{ik}(\underline{x}, t')\, \mathrm{d}t',
\end{aligned}
\right\}
\tag{8.1}
$$

where $T_{ik}$ is the extra stress tensor; $d_{ik}$ the rate-of-strain tensor; $p$ is an arbitrary isotropic pressure; $\delta_{ik}$ is the Kronecker delta, which takes the value zero for $i \neq k$

---

\* Coleman and Noll et al. define a principle called "the principle of material objectivity", which requires that the frame indifference of *principle I* must also apply to a time-dependent frame (see, for example, Truesdell and Noll 1965, Astarita and Marrucci 1974). In other words, it is *principles I* and *II* combined.

\*\* In this chapter we shall be forced to use general tensor analysis. We employ the usual notation—covariant suffices are written below, contravariant suffices above, and the usual summation convention for repeated suffices is assumed. The need to distinguish between covariance and contravariance is made clear in the detailed texts on theoretical rheology (e.g. Lodge 1974). In rectangular Cartesian coordinates, we note that it is not necessary to distinguish between covariant and contravariant components. For readers meeting *tensor analysis* for the first time, we recommend a study of Foster and Nightingale (1979, pp. 1–14) Spain (1960, pp. 8–11) and Bird et al. (1987(a), pp. 597–606).

and unity for $i = k$; $t$ is the present time and $t'$ an earlier time. $\underline{x}$ stands for $x^1$, $x^2$ and $x^3$, the Cartesian coordinates of a fixed point in space. In the corresponding one-dimensional version of (8.1) given in Chapter 3 (eqn. (3.21)), the dependence on $\underline{x}$ was not made explicit, but it was implied. The reason for this was that the strains for which the linear theory applied were so small that the particles occupied *essentially* the same position in space throughout the deformation.

If we attempt to use (8.1) under all conditions of motion and stress, we would arrive at the unphysical conclusion that the stress in the particle which is at the point $\underline{x}$ in space at the present time $t$ is determined by the history of the rate of strain in all the particles which were at the same point at previous times $t'$.

One simple (but incorrect!) way around this problem is to introduce the so-called displacement functions $x''$ (defined such that $x''$ ($i = 1, 2, 3$) is the position at time $t'$ of the element that is instantaneously at $x^i$ at time $t$) and to write

$$T_{ik}(\underline{x}, t) = 2 \int_{-\infty}^{t} \phi(t - t') d_{ik}(\underline{x}', t') \, dt', \tag{8.2}$$

where $d_{ik}$ now relates to the position $\underline{x}'$. This equation is certainly in sympathy with *principle III* since it relates "memory" to "particles" rather than "points in space", but eqns. (8.2) now equate a tensor at the point $\underline{x}$ with a tensor at the point $\underline{x}'$ and this violates *principle I* so that the satisfaction of the formulation principles (and the above discussion refers to only two of them) is clearly nontrivial.

The corresponding *differential* equations of linear viscoelasticity are also invalid under general conditions of motion and stress. Take, for example, the three-dimensional form of the simple Maxwell model given by

$$T_{ik} + \lambda \frac{\partial}{\partial t} T_{ik} = 2\eta_0 d_{ik}. \tag{8.3}$$

The partial derivative, by definition, refers to the way the stress variable is changing with time *at a particular point in space* and, at the very least, this must be replaced by the convected (Lagrangian) time derivative $D/Dt$ of hydrodynamics, in order to accommodate changes in a fluid element, where

$$\frac{D}{Dt} = \frac{\partial}{\partial t} + v^m \frac{\partial}{\partial x^m}, \tag{8.4}$$

$v^m$ being the velocity vector. However, yet again, the situation is not that simple and the correct application of *principle II* requires time derivatives of greater complexity than (8.4). These will be introduced later.

*Principles I* to *III* are of fundamental importance in the formulation of rheological equations of state; however, we would also wish such equations to satisfy two further principles:

*Principle IV.* In the case of elastic liquids, the deformation history of a material element in the distant past must be expected to have a weaker influence on the current stress response than the deformation history of the recent past. This is known as "the principle of fading memory".

*Principle V.* The equations must be consistent with thermodynamic principles. Astarita and Marrucci (1974, p. 52) show that there are pitfalls in developing a purely mechanical rheological theory without due regard being paid to thermodynamics. For instance, the principle of "positive dissipation" can provide useful constraints on constitutive equations. At the same time, much of the literature which attempts to apply in a *general* way thermodynamics to continuum mechanics has not been too successful and much of it is controversial.

## 8.3 Successful applications of the formulation principles

In his classic 1950 paper, Oldroyd sought to satisfy the basic principles of formulation by introducing a convected coordinate system $\xi^j$ embedded in the material and deforming continuously with it, so that a material element which is at $\xi^j$ at one instant is at the same point (with respect to the convected coordinate system) at every other time. Oldroyd argued that, provided one works in a tensorially-consistent manner in the convected coordinate system, the basic *principles I–III* are automatically satisfied, since using $(\xi^1, \xi^2, \xi^3, t')$ as independent variables essentially satisfies the need to concentrate on fluid elements; thus moving away from the more usual 'Eulerian' preoccupation with fixed points in space. Further, the $\xi^j$ coordinate system is unaffected by absolute (rigid body) motion in space and *principle II* is satisfied automatically (provided, of course, that operations like time differentiation and integration do not introduce any unwanted dependence on absolute motion in space). In this connection we note that a time derivative $D/Dt$ holding convected coordinates constant is a convenient and valid differential operator.

The metric tensor $\gamma_{jl}(\xi, t')$ of the $\xi^j$ system was taken by Oldroyd as the fundamental deformation variable *, since, in the usual definition,

$$[ds(t')]^2 = \gamma_{jl}(\xi, t')\, d\xi^j\, d\xi^l, \tag{8.5}$$

and $\gamma_{jl}(\xi, t')$ clearly provides a convenient measure of the distance $ds(t')$ between the parts of the arbitrary element at $\xi^j$.

Oldroyd (1950) showed how his general theory could be used by means of some simple examples. One example was the so-called liquid A, which obeyed the

---

* The contravariant metric tensor $\gamma^{jl}$ can also be used for this purpose (see, for example, Truesdell 1958, White 1964).

differential constitutive equation:

$$\tau_{jl} + \lambda_1 \frac{D}{Dt} \tau_{jl} = \eta_0 \left( \frac{D\gamma_{jl}}{Dt} + \lambda_2 \frac{D^2 \gamma_{jl}}{Dt^2} \right), \tag{8.6}$$

where $\tau_{jl}(\xi, t)$ is the extra stress tensor in the convected coordinate system * and $\eta_0$, $\lambda_1$ and $\lambda_2$ are material constants. Another example was the integral model called liquid A′:

$$\tau_{jl}(\xi, t) = \int_{-\infty}^{t} \phi(t - t') \frac{D\gamma_{jl}}{Dt'}(\xi, t') \, dt'. \tag{8.7}$$

Having written equations in the convected coordinate system $\xi^j$, Oldroyd showed how to transform equations like (8.6) and (8.7) into the fixed laboratory coordinate system $x^i$. The relevant transformation rules for so doing were given by Oldroyd (1950). For example, the $D/Dt$ time derivative of the differential models in convected coordinates must be replaced by the "codeformational" derivative $\delta/\delta t$, where, for a symmetric covariant tensor $b_{ik}$, we have

$$\frac{\delta b_{ik}}{\delta t} = \frac{\partial b_{ik}}{\partial t} + v^m \frac{\partial b_{ik}}{\partial x^m} + \frac{\partial v^m}{\partial x^i} b_{mk} + \frac{\partial v^m}{\partial x^k} b_{im}, \tag{8.8}$$

and for a symmetric contravariant tensor $b^{ik}$, the relevant form is

$$\frac{\delta b^{ik}}{\delta t} = \frac{\partial b^{ik}}{\partial t} + v^m \frac{\partial b^{ik}}{\partial x^m} - \frac{\partial v^i}{\partial x^m} b^{mk} - \frac{\partial v^k}{\partial x^m} b^{im}. \tag{8.9}$$

The derivative $\delta/\delta t$ in eqns. (8.8) is often called the lower convected derivative and that given in eqns. (8.9) the upper convected derivative. The symbol $\triangle$ over the tensor being differentiated is usually employed for the lower convected derivative and the symbol $\triangledown$ for the upper convected derivative.

Oldroyd (1950) also showed that terms like $D\gamma_{jl}(\xi, t')/Dt'$ in integral models like (8.7) need to be replaced by their so-called "Eulerian fixed components". For example, $\gamma_{jl}(\xi, t')$ has to be replaced in the fixed coordinate system $x^i$ by $G_{ik}$, where

$$G_{ik} = \frac{\partial x'^m}{\partial x^i} \frac{\partial x''^r}{\partial x^k} g_{mr}(x'), \tag{8.10}$$

---

* In his development, Oldroyd used, for consistency and convenience, Greek letters for tensor variables in the convected $\xi^j$ system and Roman letters for tensor variables in the fixed $x^i$ system.

$g_{ik}$ being the metric tensor of the $x^i$ coordinate system. When the fixed coordinate system is Cartesian, we have $g_{mr} = \delta_{mr}$, the Kronecker delta, and hence

$$G_{ik} = \frac{\partial x'^m}{\partial x^i} \frac{\partial x'^m}{\partial x^k}. \tag{8.11}$$

The equivalent stress tensor in fixed laboratory coordinates for liquid A' is

$$T_{ik}(\underline{x}, t) = 2 \int_{-\infty}^{t} \phi(t - t') \frac{\partial x'^m}{\partial x^i} \frac{\partial x''^r}{\partial x^k} d_{mr}(\underline{x}', t') \, dt'. \tag{8.12}$$

We remark that the contravariant equivalents of (8.6) and (8.7) are called liquid B and B', respectively, and are given in fixed coordinates by

$$T^{ik} + \lambda_1 \frac{\delta T^{ik}}{\delta t} = 2\eta_0 \left[ d^{ik} + \lambda_2 \frac{\delta d^{ik}}{\delta t} \right] \tag{8.13}$$

and

$$T^{ik}(\underline{x}, t) = 2 \int_{-\infty}^{t} \phi(t - t') \frac{\partial x^i}{\partial x'^m} \frac{\partial x^k}{\partial x''^r} d^{mr}(\underline{x}', t') \, dt', \tag{8.14}$$

respectively.

Oldroyd introduced models like the A and B series to illustrate the general theory, but their introduction gave the impression to some workers in the field that his formulation work was in some sense not completely general. This is now acknowledged to be a mistaken impression, but at the same time it is interesting to note that simple models like liquid B and liquid B' have figured prominently in modern developments in the numerical simulation of non-Newtonian flow (see, for example, Crochet et al. 1984). Some of the rheometrical consequences of Oldroyd-type models are given later in Table 8.3.

For an updated version of Oldroyd's work on formulation, the reader is referred to the review article published posthumously in a commemorative volume (Oldroyd 1984).

The application of the formulation principles in the work of Coleman and Noll et al. takes a somewhat different path. Some general hypothesis is made on the relationship between stress and deformation and the formulation principles, applied within a Cartesian coordinate framework, are then used either to supply the resulting equations or at least to provide constraints on them. We shall refer to this procedure as "the formal approach".

Occurring quite naturally in the development of Coleman and Noll et al. is the Eulerian fixed component equivalent of the tensor $\gamma_{jl}$, given in Cartesian coordinates by (8.11) and called the Cauchy–Green tensor. Also of importance is the series expansion of $G_{ik}$ in terms of the so-called Rivlin–Ericksen (1955) tensors $A_{ik}^{(n)}$:

$$G_{ik}(\underline{x}, t, t') = \sum_{j=0}^{\infty} \frac{(t-t')^j(-1)^j}{j!} A_{ik}^{(j)}. \qquad (8.15)$$

The Rivlin–Ericksen tensors are related to the (lower convected) time derivative of Oldroyd through

$$A_{ik}^{(n)} = 2 \frac{\delta^{n-1} d_{ik}}{\delta t^{n-1}}. \qquad (8.16)$$

In a later development, White and Metzner (1963) derived the contravariant equivalents of the Rivlin–Ericksen tensors involving the Finger tensor $F^{ik}$ in place of the Cauchy–Green tensor $G_{ik}$, where, in Cartesian coordinates

$$F^{ik} = \frac{\partial x^i}{\partial x'^m} \frac{\partial x^k}{\partial x'^m}. \qquad (8.17)$$

The resulting tensors have become known as the White–Metzner tensors and have an obvious counterpart to (8.16) with the upper convected Oldroyd derivative given by eqns. (8.9) replacing that in eqns. (8.16) (see, for example, Walters and Waterhouse 1977).

We may conclude that the basic framework of Oldroyd is matched in the developments of Coleman, Noll et al. and any differences in the application of the various techniques result from the particular outlook adopted rather than from any fundamental disagreement.

The variables and operations needed to construct rheological equations of state are now known and it is simply a matter of employing and applying these within the context of certain constitutive proposals. In these proposals there may be a preoccupation with generality or, alternatively, a search for simplicity. A compromise between the two is also a possibility.

If our concern is with generality, we may write

$$T_{ik}(t) = \mathop{\mathscr{F}}_{-\infty}^{t} \left[ G_{jl}(t') \right], \qquad (8.18)$$

which expresses mathematically the requirement that the stress at time $t$ is determined in a very general way by the history of the deformation. $\mathscr{F}$ is called a tensor-valued functional, and in the equation, the stress at time $t$ is to be viewed as a function of the deformation measure, which is itself a function of the time variable $t'$ with $-\infty < t' \leqslant t$. The *integrals* in (8.12) and (8.14) are (simple) examples of a functional.

When certain formal requirements of *functional analysis* are added to eqns. (8.18), we obtain the so called "simple fluid" of Coleman and Noll, which has had an influential impact on constitutive theory (see, for example, Truesdell and Noll 1965).

On account of their generality, eqns. (8.18) have limited predictive utility in non-Newtonian fluid mechanics and, not surprisingly, simpler equations have been sought. These arise from three distinct approaches:

*I*. One may relax the complete universality embodied in eqns. (8.18) but still make constitutive assumptions of some generality. Such developments are considered in §8.4.

*II*. One may consider approximations arising from simplifications in the flow so that $G_{jl}$ in eqns. (8.18) has a relatively simple form. These approximations lead to *general* equations of state for *restricted* classes of flow. They are discussed in §8.5.

*III*. One may consider special (usually very simple) choices of the functional $\mathscr{F}$. These lead to *particular* equations, which are nevertheless valid under *all* conditions of motion and stress. Examples of this sort are considered in detail in §8.6.

## 8.4 Some general constitutive equations

The formal approach was used by Reiner (1945) and Rivlin (1948) in a search for the most general constitutive equations for *inelastic* non-Newtonian fluids. The resulting model, which has become known as the Reiner–Rivlin model, has constitutive equations of the form

$$T_{ik} = 2\eta(I_2, I_3)d_{ik} + 4\zeta(I_2, I_3)d_i^j d_{jk}, \tag{8.19}$$

where $I_2$ and $I_3$ are the two non-zero invariants of the strain-rate tensor $d_{ik}$.

The behaviour of the Reiner–Rivlin model in a steady simple-shear flow can be easily determined. Surprisingly for an inelastic model, it predicts normal stresses. However the resulting normal stress distribution (viz. $N_1 = 0$, $N_2 \neq 0$) is not of a form which has been found in any real non-Newtonian fluid. Consequently, any normal stress differences found experimentally in a steady shear flow can be viewed as manifestations of *viscoelastic* behaviour (cf. Chapter 4).

A simplified version of the Reiner–Rivlin fluid given by

$$T_{ik} = 2\eta(I_2)d_{ik}, \tag{8.20}$$

where

$$I_2 = 2(d_i^k d_k^i)^{1/2}, \tag{8.21}$$

has been given significant prominence in the rheological literature. It is familiarly known as the "generalized Newtonian model". The form of the invariant $I_2$ given in

(8.21) is chosen such that it collapses to the shear rate $\dot{\gamma}$ in a steady simple-shear flow.

The generalized Newtonian model can account for variable viscosity effects, through the function $\eta(I_2)$, but not normal stress differences. Therefore, it has an obvious application to *fluids* which show significant viscosity variation with shear rate, but negligibly small normal stress differences, and also to *flow situations* where variable viscosity is the dominant influence (even though normal stress differences may be exhibited by the fluids under test).

In an influential development, Rivlin and Ericksen (1955) used the formal approach to derive constitutive equations based on the general proposition that the stress is a function of the velocity gradients, acceleration gradients... $(n-1)$th acceleration gradients. The resulting Rivlin–Ericksen fluid has equations of state of the form

$$T_{ik} = f\left[ A_{jl}^{(1)}, A_{jl}^{(2)} \ldots A_{jl}^{(n)} \right],$$  (8.22)

where $f$ is a function of the Rivlin–Ericksen tensors introduced in (8.15). Useful constraints on the form of the function $f$ have been found using routine matrix theory (see, for example, Truesdell and Noll 1965).

In a series of papers, Green, Rivlin and Spencer (1957, 1959, 1960) developed integral forms of eqns. (8.18), the lower-order approximations being essentially the same as the integral equations discussed in §8.5 (cf. eqns. (8.26)–(8.28)). The so-called Green–Rivlin fluids can be thought of as arising from a procedure analogous to the Taylor-series expansion of an analytic function (cf. Pipkin 1966) or, alternatively, from a direct application of the Stone–Weierstrass theorem (cf. Chacon and Rivlin 1964).

## 8.5 Constitutive equations for restricted classes of flows

There is no doubt that the simple fluid of Coleman and Noll has been the most influential application of the formal approach. The resulting equation is (8.18), with a suitably chosen function space and accompanying norm. In the original development, the function space(s) chosen by Coleman and Noll had certain limitations which were highlighted by Oldroyd (1965), who argued that the Newtonian fluid was not a special case of the simple fluid except in the limit of very slow flow. This, and related objections, have been overcome in the more recent work of Saut and Joseph (1983).

Notwithstanding the original limitations mentioned above, the basic simple fluid hypothesis was studied to good effect by Coleman and Noll, who developed simplified constitutive equations for special classes of flows. Their most influential contribution applies to so-called "slow flow", which to be precise requires the flow not only to be slow but also "slowly varying" *.

The Coleman and Noll (1960) work on the slow flow of fluids with fading memory leads to a set of approximate equations (ordered by some convenient measure of "speed of flow"), the first three being expressible in the form

$$T_{ik} = \alpha_1 A_{ik}^{(1)}, \tag{8.23}$$

$$T_{ik} = \alpha_1 A_{ik}^{(1)} + \alpha_2 A_{ik}^{(2)} + \alpha_3 A_i^{(1)j} A_{jk}^{(1)}, \tag{8.24}$$

$$T_{ik} = \alpha_1 A_{ik}^{(1)} + \alpha_2 A_{ik}^{(2)} + \alpha_3 A_i^{(1)j} A_{jk}^{(1)} + \beta_1 A_j^{(1)l} A_l^{(1)j} A_{ik}^{(1)}$$

$$+ \beta_2 A_{ik}^{(3)} + \alpha_5 \left( A_i^{(1)j} A_{jk}^{(2)} + A_i^{(2)j} A_{jk}^{(1)} \right). \tag{8.25}$$

where $\alpha_1$, $\beta_1$ etc. are all material constants. Equations (8.23) is the Newtonian model. Equations (8.24), called the second-order model, have been used extensively in modern non-Newtonian fluid mechanics.

The slow-flow development is often referred to as the "retarded-motion expansion" and the resulting equations as the "hierarchy equations" of Coleman and Noll. Equations (8.23)–(8.25) are important because they provide convenient equations for *all* simple fluids provided the flow is sufficiently slow.

Another type of approximation may be obtained from the formal approach for the case when the "deformation is small". (Such a situation exists, for example, in the case of small-amplitude oscillatory shear flow.) When certain formal "smoothness" assumptions are made, the approximation leads to integral constitutive equations of the form (Coleman and Noll 1961, Pipkin 1964)

$$T_{ik} = \int_0^\infty M_1(s) G_{ik}(s) \, ds, \tag{8.26}$$

$$T_{ik} = \int_0^\infty M_1(s) G_{ik}(s) \, ds + \int_0^\infty \int_0^\infty M_2(s_1, s_2) G_i^j(s_1) G_{jk}(s_2) \, ds_1 \, ds_2, \tag{8.27}$$

$$T_{ik} = \int_0^\infty M_1(s) G_{ik}(s) \, ds + \int_0^\infty \int_0^\infty M_2(s_1, s_2) G_i^j(s_1) G_{jk}(s_2) \, ds_1 \, ds_2$$

$$+ \int_0^\infty \int_0^\infty \int_0^\infty \left( M_1(s_1, s_2, s_3) G_i^j(s_2) G_j^l(s_3) G_{ik}(s_1) \right.$$

$$+ \left. M_4(s_1, s_2, s_3) G_i^j(s_1) G_j^l(s_2) G_{lk}(s_3) \right) \, ds_1 \, ds_2 \, ds_3, \tag{8.28}$$

---

* This is an important observation for some flow situations. For example, on account of the no-slip hypothesis, flow near a reentrant corner may be regarded as "slow" but in no sense can such a flow be regarded as "slowly-varying", so that the Coleman and Noll development for the flow of fluids with fading memory does not apply to such situations (see, for example, Crochet et al. 1984).

where $s = t - t'$ is the time lapse and, from the symmetry of the stress tensor, the kernel functions must satisfy

$$M_2(s_1, s_2) = M_2(s_2, s_1), \quad M_4(s_1, s_2, s_3) = M_4(s_3, s_2, s_1). \tag{8.29}$$

Equations (8.26) are called the equations of finite linear viscoelasticity *, whilst (8.27) are called the equations of second-order viscoelasticity and so on. The small-deformation development of Coleman and Noll has much in common with the integral expansions of Green, Rivlin and Spencer (1957, 1959, 1960). Crochet et al. (1984) advocate care in the use of the integral expansions since their range of applicability is not as wide as might be anticipated.

Many of the flow problems which are tractable by analytic methods fall into the category of "nearly viscometric flows", which are flows that are close to viscometric flows like Poiseuille or Couette flow; the "closeness" can be defined in a precise mathematical way. Pipkin and Owen (1967) have addressed the possibility of obtaining constitutive equations for this restricted class of flows. They conclude that, in an integral formulation, thirteen independent kernel functions are required.

So-called "motions with constant stretch history" have been studied by numerous theoretical rheologists, and the associated constitutive equations have been derived. The subject is covered in the books by Huilgol (1975), Lodge (1974) and Dealy (1982).

## 8.6 Simple constitutive equations of the Oldroyd/Maxwell type

The developments discussed in §8.4 and §8.5 must be viewed as important contributions to the subject, but most of the associated equations are of limited utility in the solution of practical flow problems, either on account of their complexity or their limited range of applicability. Accordingly, numerous attempts have been made to develop relatively simple constitutive equations with predictive capability. The form of these equations may be guided by a knowledge of the fluid's microstructure (cf. §6.8) or by the requirement that they must be able to simulate real behaviour in simple (rheometrical) flow situations. For example, the popular Oldroyd models arose originally from a desire to generalize (for all conditions of motion and stress) relatively simple linear equations like the Jeffreys model (eqn. (3.15)) which were known to be useful approximations for very dilute suspensions and emulsions under conditions of small strain (see, for example, Fröhlich and Sack 1946, Oldroyd 1953, Oldroyd 1958).

It must be admitted that to model microstructure in any complete way would require prohibitive detail and some compromise is needed between capturing the

---

* Employing an integration by parts, it is possible to show that (8.26) is equivalent to liquid A' (eqns. (8.12)).

known complexity of the physics and generating equations with predictive capability.

A further factor of importance in the choice of constitutive model is the application in mind. For example, it is more important for the model to represent the *extensional-viscosity* characteristics (rather than, say, the normal-stress differences) if the model is to be employed in a fibre spinning problem.

In summary, simple constitutive models have to satisfy, *if possible*, the following:

*(i)* they must satisfy the formulation principles discussed in §8.2. This is clearly not an optional requirement. The Oldroyd approach is ideally suited for this purpose;

*(ii)* they should reflect the physics of the microstructure;

*(iii)* they should be able to simulate the behaviour of the fluid in simple flows like steady simple shear, oscillatory shear and extensional flow;

*(iv)* they should have regard to the application in mind.

Given these constraints and the plethora of possibilities, it is not appropriate for us to favour one model at the expense of others, especially in view of the fact that history suggests that the popularity of a given model is often ephemeral. Rather, we list in tabular form many of the popular differential constitutive models which have appeared in the literature and can be viewed as having predictive capacity (see Tables 8.1 and 8.2). These can all be regarded as special cases of the general canonical forms *:

$$T_{ik} = T_{ik}^{(1)} + T_{ik}^{(2)}, \tag{8.30}$$

in which the terms on the right hand side are given by

$$\exp\left(\epsilon\frac{\lambda_1}{\eta_1}T_{mm}^{(1)}\right)T_{ik}^{(1)} + \alpha\frac{\lambda_1}{\eta_1}T_{ij}^{(1)}T_{jk}^{(1)} + \lambda_1\overset{\square}{T}_{ik}^{(1)} = 2\eta_1 d_{ik}, \tag{8.31}$$

$$T_{ik}^{(2)} = 2\eta_2 d_{ik}, \tag{8.32}$$

where, unless otherwise stated, $\lambda_1$, $\eta_1$, $\eta_2$, $\epsilon$, and $\alpha$ are all material constants and the derivative $\square$ is given by **

$$\overset{\square}{S}_{ik} = \left(1 - \frac{a}{2}\right)\overset{\triangledown}{S}_{ik} + \frac{a}{2}\overset{\triangle}{S}_{ik}, \tag{8.33}$$

where $a$ is a scalar parameter.

---

\* For convenience, the models are expressed in a form appropriate to a rectangular Cartesian coordinate system.

\*\* We recommend Crochet et al. (1984, Chapter 2) and Giesekus (1984) for a fuller discussion of the various time-derivatives of continuum mechanics.

TABLE 8.1
Special cases of the general canonical form of constitutive equation (eqns. 8.30–8.33)

| Model | $\epsilon$ | $\alpha$ | $a$ | $\lambda_1$ ($\geqslant 0$) | $\eta_1$ ($\geqslant 0$) | $\eta_2$ ($\geqslant 0$) |
|---|---|---|---|---|---|---|
| Giesekus (1982) | 0 | $\alpha$ | 0 | $\lambda_1$ | $\eta_1$ | 0 |
| Phan-Thien–Tanner (1977) | $\epsilon$ | 0 | $a$ | $\lambda_1$ | $\eta_1$ | 0 |
| Phan-Thien–Tanner B (Phan-Thien 1984) | 0 | 0 | $a$ | $\lambda_1$ | $\eta_1(I_2)$ | 0 |
| Johnson–Segalman (1977) | 0 | 0 | $a$ | $\lambda_1$ | $\eta_1$ | 0 |
| White–Metzner (1963) | 0 | 0 | 0 | $\lambda_1(I_2)$ | $\eta_1(I_2)$ | 0 |
| Oldroyd B | 0 | 0 | 0 | $\lambda_1$ | $\eta_1$ | $\eta_2$ |
| Corotational Oldroyd (Oldroyd 1958) | 0 | 0 | 1 | $\lambda_1$ | $\eta_1$ | $\eta_2$ |
| Upper convected Maxwell | 0 | 0 | 0 | $\lambda_1$ | $\eta_1$ | 0 |
| Second-order model | | | | Not applicable in this form | | |
| Leonov (1987) | 0 | $\frac{1}{2}$ | 0 | $\lambda_1$ | $\eta_1$ | 0 |

We may also use the alternative canonical form

$$\exp\left(\epsilon\frac{\lambda_1}{\eta_1}T_{mm}\right)T_{ik} + \alpha\frac{\lambda_1}{\eta_1}T_{ij}T_{jk} + \lambda_1\overset{\square}{T}_{ik} = 2\eta_1\left(d_{ik} + \lambda_2\overset{\square}{d}_{ik}\right), \tag{8.34}$$

where we have now omitted the $\eta_2$ contribution and essentially replaced it by a retardation time $\lambda_2$.

TABLE 8.2
Special cases of the general canonical form of constitutive equation (eqn. 8.34)

| Model | $\epsilon$ | $\alpha$ | $a$ | $\lambda_1$ ($\geqslant 0$) | $\lambda_2$ | $\eta_1$ ($> 0$) |
|---|---|---|---|---|---|---|
| Giesekus (1982) | 0 | $\alpha$ | 0 | $\lambda_1$ | 0 | $\eta_1$ |
| Phan-Thien–Tanner (1977) | $\epsilon$ | 0 | $a$ | $\lambda_1$ | 0 | $\eta_1$ |
| Phan-Thien–Tanner B (Phan-Thien 1984) | 0 | 0 | $a$ | $\lambda_1$ | $\lambda_2$ | $\eta_1(I_2)$ |
| Johnson–Segalman (1977) | 0 | 0 | $a$ | $\lambda_1$ | 0 | $\eta_1$ |
| White–Metzner (1963) | 0 | 0 | 0 | $\lambda_1(I_2)$ | 0 | $\eta_1(I_2)$ |
| Oldroyd B | 0 | 0 | 0 | $\lambda_1$ | $> 0$ | $\eta_1$ |
| Corotational Oldroyd (Oldroyd 1958) | 0 | 0 | 1 | $\lambda_1$ | $> 0$ | $\eta_1$ |
| Upper convected Maxwell | 0 | 0 | 0 | $\lambda_1$ | 0 | $\eta_1$ |
| Second-order model | 0 | 0 | 0 | 0 | $< 0$ | $\eta_1$ |
| Leonov (1987) | 0 | $\frac{1}{2}$ | 0 | $\lambda_1$ | 0 | $\eta_1$ |

TABLE 8.3
Rheometrical forms derived from some of the models in Tables 8.1 and 8.2

| Model | $\eta(\dot{\gamma})$ | $N_1(\dot{\gamma})$ | $N_2(\dot{\gamma})$ | $\eta_E(\dot{\epsilon})$ |
|---|---|---|---|---|
| Oldroyd B | $\eta_1 + \eta_2$ | $2\eta_1\lambda_1\dot{\gamma}^2$ | 0 | $\dfrac{2\eta_1}{1-2\lambda_1\dot{\epsilon}} + \dfrac{\eta_1}{1+\lambda_1\dot{\epsilon}}$ $+3\eta_2$ |
| Corotational Oldroyd | $\dfrac{\eta_1}{1+\lambda_1^2\dot{\gamma}^2} + \eta_2$ | $\dfrac{2\lambda_1\eta_1\dot{\gamma}^2}{1+2\lambda_1^2\dot{\gamma}^2}$ | $-\dfrac{N_1(\dot{\gamma})}{2}$ | $3(\eta_1+\eta_2)$ |
| Johnson–Segalman (1977) | $\dfrac{\eta_1}{1+2a\left(1-\dfrac{a}{2}\right)\lambda_1^2\dot{\gamma}^2}$ | $\dfrac{2\eta_1\lambda_1\dot{\gamma}^2}{1+2a\left(1-\dfrac{a}{2}\right)\lambda_1^2\dot{\gamma}^2}$ | $-\dfrac{a}{2}N_1(\dot{\gamma})$ | $\dfrac{2\eta_1}{1-2(1-a)\lambda_1\dot{\epsilon}}$ $+\dfrac{\eta_1}{1+(1-a)\lambda_1\dot{\epsilon}}$ |
| White–Metzner (1963) | $\eta_1(\dot{\gamma})$ | $2\eta(\dot{\gamma})\lambda_1(\dot{\gamma})\dot{\gamma}^2$ | 0 | $\dfrac{2\eta_1(\sqrt{3}\,\dot{\epsilon})}{1-2\lambda_1(\sqrt{3}\,\dot{\epsilon})\dot{\epsilon}}$ $+\dfrac{\eta_1(\sqrt{3}\,\dot{\epsilon})}{1+\lambda_1(\sqrt{3}\,\dot{\epsilon})\dot{\epsilon}}$ |
| Phan-Thien–Tanner B (Phan-Thien 1984) | $\dfrac{\eta_1(\dot{\gamma})}{1+2a\left(1-\dfrac{a}{2}\right)\lambda_1^2\dot{\gamma}^2}$ | $\dfrac{2\eta_1(\dot{\gamma})\lambda_1\dot{\gamma}^2}{1+2a\left(1-\dfrac{a}{2}\right)\lambda_1^2\dot{\gamma}^2}$ | $-\dfrac{a}{2}N_1(\dot{\gamma})$ | $\dfrac{2\eta_1(\sqrt{3}\,\dot{\epsilon})}{1-2(1-a)\lambda_1\dot{\epsilon}}$ $+\dfrac{\eta_1(\sqrt{3}\,\dot{\epsilon})}{1+(1-a)\lambda_1\dot{\epsilon}}$ |
| Giesekus (1982) | | Consult the reference for detailed expressions | | |

Equations (8.32), and by implication the term in $\lambda_2$ in (8.34), can be viewed as a "Newtonian dashpot" contribution, either introduced to reflect the solvent contribution in liquids like polymer solutions or to ensure that the shear stress in a steady simple shear flow is a monotonic increasing function of shear rate. Some of the so-called Maxwell models (with $\eta_2 = 0$ or $\lambda_2 = 0$) suffer from the problem of a stress maximum unless $a = 0$ or 2.

For convenience, we list in Table 8.3 the main rheometrical functions derived from many of the models introduced in Tables 8.1 and 8.2.

Not surprisingly, there have been similar developments involving *integral* equations instead of the implicit differential models discussed above. These range from the comparative simplicity of the Lodge (1956) rubber-like liquid (which is essentially equivalent to liquid B′) and more or less stop at the complexity provided by the so-called KBKZ model (cf. Bernstein et al. 1963), with constitutive equations which, in Cartesian coordinates, are essentially given by

$$T_{ik} = \int_{-\infty}^{t} \left[ \phi_1(I_1, I_2, t-t')G_{ik}(t') + \phi_2(I_1, I_2, t-t')G_{ij}(t')G_{jk}(t') \right] \, dt',$$

$$(8.35)$$

where $I_1$ and $I_2$ are now the two non-zero invariants of $G_{ik}$. We remark that the Doi–Edwards model discussed in §6.8.5 (with the added "independent-alignment" assumption) leads to an equation of the KBKZ type (see, for example, Doi and Edwards 1986, Marrucci 1986).

## 8.7 Solution of flow problems

We now consider the application of the work of the previous sections to the solution of non-Newtonian flow problems. To facilitate this, it is helpful to attempt a flow classification (cf. Crochet et al. 1984, Chapter 3).

*I*. When the flow is "slow", the choice of constitutive model is self-evident (i.e. one of the hierarchy models (8.23)–(8.25)) and there is no merit whatsoever in employing any of the more complicated implicit differential or integral models discussed in the previous section (cf. Walters 1979). Flow problems in the case of slow flow invariably resolve themselves into perturbation analyses with "speed of flow" as the relevant perturbation parameter.

*II*. When the flow is dominated by the shear viscosity, the generalized Newtonian model (8.20) can be employed.

*III*. Many of the flow problems which have been solved successfully using analytic techniques fall into the category of "nearly viscometric flows". Linear stability analyses, flow caused by rotating bodies and various pipe flows can be placed in the category of nearly viscometric flows. We have already indicated that the general description of such flows is of prohibitive complexity and approximate equations have been employed in existing analyses. These become perturbation problems using the basic viscometric flow as the primary flow and a convenient (geometrical, flow or continuum) parameter as the perturbation variable.

*IV*. The advent of powerful digital computers has seen interest in non-Newtonian fluid mechanics moving towards the solution of complex flow problems for highly elastic liquids: situations which are of practical importance. Differential and integral equations at all levels of complexity are being employed in this expanding research field and it is probably true to say that, within reason, there are now few restrictions on the amount of detail that can be handled in the constitutive equation employed. The subject is covered in detail in the text by Crochet et al. (1984). The so called "high Weissenberg-number problem" which restricted all early work in the field is discussed in that text. However, we remark with interest that the recent work of Crochet and his collaborators (cf. Marchal and Crochet 1987) has not been so hampered by the high Weissenberg-number problem and the resulting numerical simulations are valid for conditions of practical importance where major changes in flow characteristics are observed. Any current discrepancies between theory and experiment, and there are still some, can no longer be attributed (solely) to the high Weissenberg-number problem. Attention must now be focussed on other problems, viz.:

(*i*) the possibility of three-dimensional flow characteristics occurring in seemingly two-dimensional flows;

(*ii*) the possibility of "bifurcation" and lack of uniqueness in complex flows of highly elastic liquids;

(*iii*) the inadequacies of the constitutive equations in current use for very complex materials;

(*iv*) the incorrect numerical treatment of flow near reentrant corners and possibly also the incorrect numerical treatment of the extra constitutive difficulties associated with long-range fluid memory.

The field of the numerical simulation of non-Newtonian flow is developing rapidly and a constant update on current literature is recommended in this area.

# GLOSSARY OF RHEOLOGICAL TERMS

This glossary is based mainly on the British Standard of the same title and numbered BS 5168:1975. It differs from the British Standard in that it is not intended to be comprehensive but limited to the terms which are most relevant to the present book. All quantitative terms have been given their SI units and symbols where this is feasible. The symbols are those used in this book and are therefore recommended; they include many which are recommended by the U.S. Society of Rheology and which were published in the Journal of Rheology (1984) $\underline{28}$, 181–195.

| | |
|---|---|
| Anti-thixotropy | An increase of the apparent viscosity under constant shear stress/rate followed by a gradual recovery when the stress or shear rate is removed. The effect is time-dependent (see negative thixotropy and rheopexy). |
| Apparent viscosity | The shear stress divided by rate of shear when this quotient is dependent on rate of shear. Also called viscosity and shear viscosity. $\eta$ Pa.s. |
| Bingham model | A model with the behaviour of an elastic solid up to the yield stress; above the yield stress, the rate of shear is directly proportional to the shear stress minus the yield stress. |
| Biorheology | The study of the rheological behaviour of biological materials. |
| Complex (shear) compliance | The mathematical representation of a (shear) compliance as the sum of a real and an imaginary part. The real part is sometimes called storage compliance and the imaginary part loss compliance. $J^{\star}$ (for shear) $Pa^{-1}$. |

| Complex (shear) modulus | The mathematical representation of a (shear) modulus as the sum of a real and an imaginary part. The real part is sometimes called storage modulus and the imaginary part loss modulus. $G^{\star}$ (for shear) Pa. |
|---|---|
| Complex viscosity | The mathematical representation of a viscosity as the sum of a real part and an imaginary part. The real part is usually called dynamic viscosity, the imaginary part is related to the real part of the complex shear modulus. $\eta^{\star}$ Pa.s. |
| Compliance | The strain divided by the corresponding stress. $J$ (for shear) $Pa^{-1}$. |
| Consistency | A general term for the property of a material by which it resists permanent change of shape. |
| Constitutive equation | An equation relating stress, strain, time and sometimes other variables such as temperature. Also called rheological equation of state. |
| Continuum rheology | The rheology that treats a material as a continuum without explicit consideration of microstructure. Also called macrorheology and phenomenological rheology. |
| Couette flow (circular) | Simple shear flow in the annulus between two co-axial cylinders in relative rotation. |
| Couette flow (plane) | Simple shear flow between parallel plates in relative motion in their own plane. |
| Creaming | The rising of particles of the dispersed phase to the surface of a suspension. |
| Creep | The slow deformation of a material; usually measured under constant stress. |
| Dashpot | A model for Newtonian viscous flow, typically represented by a piston moving in a cylinder of liquid. |

| | |
|---|---|
| Deborah number | The ratio of a characteristic (relaxation) time of a material to a characteristic time of the relevant deformation process. |
| Deformation | A change of shape or volume or both. |
| Die swell | A post-extrusion swelling. |
| Dilatancy | (1) An increase in volume caused by shear. |
| | (2) Shear thickening (deprecated usage). |
| Dynamic modulus | Synonym of complex modulus. |
| Dynamic viscosity | (1) In classical fluid mechanics a synonym of coefficient of viscosity used to distinguish this quantity from kinematic viscosity. $\eta$ Pa.s. |
| | (2) In rheology, the quotient of the part of the stress in phase with the rate of strain divided by the rate of strain under sinusoidal conditions. $\eta'$ Pa.s. |
| Elastic(ity) | A reversible stress/strain behaviour. |
| Elastic energy | Synonym of strain energy. |
| Elastic liquid | A liquid showing elastic as well as viscous properties (see elastico-viscous, viscoelastic and memory fluid). |
| Elastic modulus | A stress divided by the corresponding elastic strain. Pa. |
| Elastico-viscous | A descriptive term for a liquid having both viscous and elastic properties. |
| Elongational viscosity | Synonym of extensional viscosity. |
| Equation of state | Synonym of constitutive equation. |

| | |
|---|---|
| Extensional viscosity | The extensional (tensile) stress divided by the rate of extension. Also called elongational viscosity and Trouton viscosity. $\eta_E$ Pa.s. |
| Extensional strain | Relative deformation in extension. $\epsilon$. |
| Extensional strain rate | The change in extensional strain per unit time. $\dot{\epsilon}$ s$^{-1}$. |
| Extra-stress tensor | The difference between the stress tensor and the isotropic pressure contribution; used for incompressible materials. $T_{ik}$ Pa. |
| Flow | A deformation, of which at least part is non-recoverable (rheological usage). |
| Flow birefringence | The optical anisotropy caused by flow. |
| Flow curve | A curve relating stress to rate of shear (cf. rheogram). |
| High elasticity | The ability of a material to undergo large elastic strains. |
| Hooke model | A model representing Hooke's law of elasticity, e.g. a spring. |
| Intrinsic viscosity | The limiting value of the reduced viscosity as the concentration approaches zero. $[\eta]$. |
| Isotropic | Having the same property in all directions. |
| Kelvin model | A mechanical model consisting of a Hooke model and Newtonian fluid model in parallel. Also called Voigt model. |
| Kinematic viscosity | In classical fluid mechanics, the dynamic viscosity divided by the density of the material. $\nu$ m$^2$ s$^{-1}$. |
| Laminar flow | Flow without turbulence. |

| | |
|---|---|
| Linear viscoelasticity | Viscoelasticity characterized by a linear relationship between stress and strain. |
| Loss angle | The phase difference between the stress and strain in an oscillatory deformation. |
| Loss compliance | The imaginary part of the complex compliance. $J''$ (for shear) $Pa^{-1}$. |
| Loss modulus | The imaginary part of the complex modulus. $G''$ (for shear) Pa. |
| Macrorheology | Synonym of continuum rheology. |
| Maxwell model | A mechanical model consisting of a Hooke model and a Newtonian fluid model in series. |
| Melt fracture | The irregular distortion of a polymer extrudate after passing through a die. |
| Memory fluid | Synonym of elastic liquid. |
| Microrheology | The rheology in which account is taken of the microstructure of materials. |
| Model | An idealized relationship of rheological behaviour expressible in mathematical, mechanical or electrical terms. |
| Modulus | In rheology, the ratio of a component of stress to a component of strain. Pa. |
| Navier–Stokes equations | The equations governing the motion of a Newtonian fluid. |
| Necking | The non-uniform local reduction of the cross-sectional area of a test piece under extension. |
| Negative thixotropy | Synonym of anti-thixotropy. |

Newtonian fluid model    A model characterized by a constant value for the quotient of the shear stress divided by the rate of shear in a simple shear flow and with zero normal stress differences (see dashpot).

Non-Newtonian fluid    Any fluid whose behaviour is not characterized by the Navier–Stokes equations.

Normal force    (1) A force acting at right angles to a specified ares. N.
(2) In rheology, a force acting at right angles to an applied shear stress. N.

Normal stress    The component of stress at right angles to the area considered. $\sigma_{11}$ Pa.

Normal stress difference    The difference between normal stress components. $N_1$ Pa.

Normal stress coefficient    A normal stress difference divided by the square of the rate of shear. $\Psi_1$ Pa s$^2$

Overshoot    The transient rise of a stress above the equilibrium value at the start up of simple shear flow.

Plastic(ity)    The capacity of a material to be moulded but also to retain its shape for a significant period under finite forces; showing flow above a yield stress.

Plastic viscosity    For a Bingham model, the excess of the shear stress over the yield stress divided by the rate of shear. $\eta_p$ Pa.s.

Poiseuille flow    Laminar flow in a pipe of circular cross section under a constant pressure gradient.

Power-law behaviour    Behaviour characterized by a linear relationship between the logarithm of the shear stress and the logarithm of the rate of shear in simple shear flow.

Pseudoplasticity    Synonym for shear thinning (usage deprecated).

| | |
|---|---|
| Rate of shear | (See shear rate). |
| Reduced viscosity | The specific viscosity per unit concentration of the solute or the dispersed phase. $m^3 \ kg^{-1}$. |
| Relative deformation | The measurement of deformation relative to a reference configuration of length, area or volume. Also called strain. |
| Relative viscosity | The ratio of the viscosity of a solution to that of the solvent or of a dispersion to that of its continuous phase (see viscosity ratio). $\eta_r$. |
| Relaxation time | The time taken for the shear stress of a fluid that obeys the Maxwell model to reduce to $1/e$ of its original equilibrium value on the cessation of steady shear flow. |
| Retardation time | The time taken for the strain in a material that obeys the Kelvin model to reduce to $1/e$ of its original equilibrium value after the removal of the stress. |
| Reynolds number | The product of a typical apparatus length and a typical fluid speed divided by the kinematic viscosity of the fluid. It expresses the ratio of the inertia forces to the viscous forces. $R_e$. |
| Rheogoniometer | A rheometer designed for the measurement of normal as well as shear components of the stress tensor. |
| Rheogram | A graph of a rheological relationship. |
| Rheological equation of state | Synonym of constitutive equation. |
| Rheology | The science of the deformation and flow of matter. |
| Rheometer | An instrument for measuring rheological properties. |
| Rheopexy | Synonym of anti-thixotropy. |

Rigidity modulus          Synonym of shear modulus.

Secondary flow            The components of flow in a plane orthogonal to the
                          main direction of flow.

Shear                     (1) The movement of a layer of material relative to
                              parallel adjacent layers.
                          (2) An abbreviation of shear strain.

Shear compliance          The elastic shear strain divided by the corresponding
                          shear stress. $J$ $Pa^{-1}$.

Shear modulus             The shear stress divided by the corresponding elastic
                          shear strain. Also known as rigidity modulus. $G$ Pa.

Shear rate                The change of shear strain per unit time. $\dot{\gamma}$ $s^{-1}$.

Shear strain              Relative deformation in shear; term often abbreviated to
                          shear. $\gamma$.

Shear stress              The component of stress parallel to (tangential to) the
                          area considered. $\sigma$ Pa.

Shear-thickening          The increase of viscosity with increasing rate of shear in
                          a steady shear flow.

Shear-thinning            The reduction of viscosity with increasing rate of shear
                          in a steady shear flow.

Shear viscosity           Synonym of apparent viscosity.

Simple shear              A shear caused by the parallel relative displacement of
                          parallel planes (see viscometric flow).

Soft solid                A descriptive term for a material exhibiting plastic
                          behaviour.

| | |
|---|---|
| Specific viscosity | The difference between the viscosity of a solution or dispersion and that of the solvent or continuous phase, divided by the viscosity of the solvent or continuous phase. $\eta_{sp}$. |
| Spinnability | The capacity of a liquid to form stable extended threads. |
| Steady flow | A flow in which the velocity at every point does not vary with time. |
| Storage compliance | That part of the (shear) strain that is in phase with the (shear) stress divided by the stress under sinusoidal conditions. $J'$ (for shear) $Pa^{-1}$. |
| Storage modulus | That part of the (shear) stress that is in phase with the (shear) strain divided by the strain under sinusoidal conditions. $G'$ (for shear) Pa. |
| Stored energy | Synonym of strain energy. |
| Strain | The measurement of deformation relative to a reference configuration of length, area or volume. Also called relative deformation. |
| Strain energy | The energy stored in a material (per unit volume) by the elastic strain. Also called elastic energy. $Jm^{-3}$. |
| Stress | A force per unit area. Pa. |
| Stress relaxation | The decrease of (shear) stress on the cessation of steady (shear) flow, usually when the stress in the original steady (shear) flow has reached equilibrium. |
| Stress tensor | A matrix of the shear stress and normal stress components representing the state of stress at a point in a body. $\sigma_{ik}$ Pa. |

Taylor number

A dimensionless group associated with viscous instabilities in circular Couette flow, the value of which depends on the kinematic viscosity and on the radii and velocities of the cylinders.

Taylor vortices

The secondary flow consisting of ring-like cell vortices associated with an instability in circular Couette flow when the Taylor number exceeds a certain value.

Tension

A force normal to the surface on which it acts and directed outwards from the body. N.

Tension-thickening

An increase in extensional viscosity with increasing rate of strain in a steady extensional flow.

Tension-thinning

A decrease in extensional viscosity with increasing rate of strain in a steady extensional flow.

Thixotropy

A decrease of the apparent viscosity under constant shear stress or shear rate, followed by a gradual recovery when the stress or shear rate is removed. The effect is time-dependent.

Time-temperature superposition

The scaling of the results of shear strain experiments carried out at different temperatures to fit onto a single curve.

Trouton ratio

The ratio of extensional to shear viscosities.

Trouton viscosity

Synonym of extensional viscosity.

Turbulence

A condition of flow in which the velocity components show random variation.

Velocity gradient

The derivative of the velocity of a fluid element with respect to a space coordinate. $s^{-1}$.

| | |
|---|---|
| Viscoelastic(ity) | Having both viscous and elastic properties. This term is sometimes restricted to solids. |
| Viscometer | An instrument for the measurement of viscosity. |
| Viscometric flow | A laminar flow which is equivalent to a steady simple-shear flow. Such a flow is determined by a maximum of three material functions: the viscosity function and two normal stress functions. |
| Viscosity | (1) Qualitatively, the property of a material to resist deformation increasingly with increasing rate of deformation.<br>(2) Quantitatively, a measure of this property, defined as the shear stress divided by the rate of shear in steady simple-shear flow. Often used synonymously with apparent viscosity. Pa.s. |
| Viscosity ratio | Synonym of relative viscosity. |
| Voigt model | Synonym of Kelvin model. |
| Weissenberg effect | An effect found in non-Newtonian fluids manifested, for example, in the climbing of the fluid up a rotating rod dipping into it. A normal stress effect. |
| Weissenberg number | The product of the relaxation time, or some other characteristic time of a material, and the rate of shear of the flow. $W_e$. |
| Yield stress | The stress corresponding to the transition from elastic to plastic deformation. $\sigma_y$ Pa. |
| Young's modulus | The extensional (tensile) stress divided by the corresponding extensional strain of an elastic material, measured in uniaxial extension. $E$ Pa. |

# REFERENCES

[1906]
EINSTEIN, A.: Eine neue Bestimmung der Molekuldimension, *Ann. Physik* 19, 289–306.

[1911]
EINSTEIN, A.: Berichtigung zu meiner Arbeit: Eine neue Bestimmung der Molekuldimension, *Ann. Physik* 34, 591–592.

[1945]
ALFREY, T.Jr.: Methods of representing the properties of viscoelastic materials, *Quart. Appl. Math.* 3, 143–150.
REINER, M.: A mathematical theory of dilatancy, *Am. J. Math.* 67, 350–362.

[1946]
FRÖHLICH, H., SACK, R.: Theory of the rheological properties of dispersions, *Proc. Roy. Soc.* A185, 415–430.
GREEN, M.S., TOBOLSKY, A.V.: A new approach to the theory of relaxing polymeric media, *J. Chem. Phys.* 14, 80–92.

[1948]
RIVLIN, R.S.: The hydrodynamics of non-Newtonian fluids, Part 1, *Proc. Roy. Soc.* A193, 260–281.

[1950]
OLDROYD, J.G.: On the formulation of rheological equations of state, *Proc. Roy. Soc.* A200, 523–541.
ROSCOE, R.: Mechanical models for the representation of viscoelastic properties, *Brit. J. Appl. Phys.* 1, 171–173.

[1953]
GROSS, B.: "*Mathematical Structure of the Theories of Viscoelasticity*", Hermann et Cie., Paris.
OLDROYD, J.G.: The elastic and viscous properties of emulsions and suspensions, *Proc. Roy. Soc.* A218, 122–132.
ROBERTS, J.E.: Pressure distribution in liquids in laminar shearing motion and comparison with predictions from various theories, in *Proceedings of the 2nd. International Congress on Rheology*, Ed. V.G.W. Harrison, Butterworths Sci. Pub., 91–98.
ROUSE, P.E.Jr.: A theory of linear viscoelastic properties of dilute solutions of coiling polymers, *J. Chem. Phys.* 21, 1272–1280.

[1954]
BUECHE, F.: Influence of rate of shear on the apparent viscosity of dilute polymer solutions and bulk polymers, *J. Chem. Phys.* 22, 1570–1576.
KRIEGER, I.M., MARON, S.H.: Direct determination of the flow curves of non-Newtonian fluids, *J. Appl. Phys.* 25, 72–75.

[1955]

RIVLIN, R.S., ERICKSEN, J.L.: Stress-deformation relations for isotropic materials, *J. Rat. Mech. Anal.* 4, 323–425.


[1956]

LODGE, A.S.: A network theory of flow birefringence and stress in concentrated polymer solutions, *Trans. Faraday Soc.* 52, 120–130.

STAVERMAN, A.J., SCHWARZL, F.R.: *"Die Physik der Hochpolymeren"* Vol 4, Ed. H.A. Stuart, Springer Verlag, 215–235.

TORDELLA, J.P.: Fracture in the extrusion of amorphous polymers through capillaries, *J. Appl. Phys.* 27, 454–458.

ZIMM, B.H.: Dynamics of polymer molecules in dilute solution: viscoelasticity, flow birefringence and dielectric loss, *J. Chem. Phys.* 24, 269–278.


[1957]

BAGLEY, E.B.: End corrections in the capillary flow of polyethylene, *J. Appl. Phys.* 28, 624–627.

GREEN, A.E., RIVLIN, R.S.: The mechanics of non-linear materials with memory, Part 1, *Arch. Rat. Mech. Anal.* 1, 1–21.


[1958]

COX, W.P., MERZ, E.H.: Correlation of dynamic and steady flow viscosities, *J. Polym. Sci.* 28, 619–622.

OLDROYD, J.G.: Non-Newtonian effects in steady motion of some idealized elastico-viscous liquids, *Proc. Roy. Soc.* A245, 278–297.

SISKO, A.W.: The flow of lubricating greases, *Ind. Eng. Chem.* 50, 1789–1792.

TRUESDELL, C.: Geometrical interpretation for the reciprocal deformation tensors, *Quart. Appl. Math.* 15, 434–435.


[1959]

GREEN, A.E., RIVLIN, R.S., SPENCER, A.J.M.: The mechanics of non-linear materials with memory, Part 2, *Arch. Rat. Mech. Anal.* 3, 82–90.

KRIEGER, I.M., DOUGHERTY, T.J.: A mechanism for non-Newtonian flow in suspensions of rigid spheres, *Trans. Soc. Rheol.* 3, 137–152.


[1960]

BRAMHALL, A.D., HUTTON, J.F.: Wall effect in the flow of lubricating greases in plunger viscometers, *Brit. J. Appl. Phys.* 11, 363–371.

COLEMAN, B.D., NOLL, W.: An approximation theorem for functionals, with applications in continuum mechanics, *Arch. Rat. Mech. Anal.* 6, 355–370.

GREEN, A.E., RIVLIN, R.S.: The mechanics of non-linear materials with memory, Part 3, *Arch. Rat. Mech. Anal.* 4, 387–404.

SPAIN, B.: *"Tensor Calculus"*, Oliver and Boyd.


[1961]

COLEMAN, B.D., NOLL, W.: Foundations of linear viscoelasticity, *Rev. Mod. Phys.* 33, 239–249.


[1962]

SEGRE, G., SILBERBERG, A.: Behaviour of macroscopic rigid spheres in Poiseuille flow. Part 1, *J. Fluid Mech.* 14, 115–135; Part 2, *J. Fluid Mech.* 14, 136–157.

**[1963]**

BERNSTEIN, B., KEARSLEY, A.E., ZAPAS, L.: A study of stress relaxation with finite strain, *Trans. Soc. Rheol.* 7, 391–410.

KOLSKY, H.: *"Stress Waves in Solids"*, Dover Publications.

van WAZER, J.R., LYONS, J.W., KIM, K.Y., COLWELL, R.E.: *"Viscosity and Flow Measurement: A Laboratory Handbook of Rheology"*, Interscience, London.

WHITE, J.L., METZNER, A.B.: Development of constitutive equations for polymeric melts and solutions, *J. Appl. Polym. Sci.* 7, 1867–1889.

**[1964]**

CHACON, R.V.S., RIVLIN, R.S., Representation theorems in the mechanics of materials with memory, *Z.A.M.P.* 15, 444–447.

LODGE, A.S.: *"Elastic Liquids"*, Academic Press.

OLDROYD, J.G.: Nonlinear stress, rate of strain relations at finite rates of shear in so-called "linear" elastico-viscous liquids, in *"Second-order effects in Elasticity, Plasticity and Fluid Dynamics"*, Ed. D. Abir, Pergamon, 520–529.

PIPKIN, A.C.: Small finite deformation of viscoelastic solids, *Rev. Mod. Phys.* 36, 1034–1041.

WHITE, J.L.: A continuum theory of nonlinear viscoelastic deformation with application to polymer processing, *J. Appl. Polym. Sci.* 8, 1129–1146.

**[1965]**

CROSS, M.M.: Rheology of non-Newtonian fluids: a new flow equation for pseudo-plastic systems, *J. Colloid Sci.* 20, 417–437.

HUTTON, J.F.: The fracture of liquids in shear: the effects of size and shape, *Proc. Roy. Soc.* A287, 222–239.

MAXWELL, B., CHARTOFF, R.P.: A polymer melt in an orthogonal rheometer, *Trans. Soc. Rheol.* 9, 41–52.

OLDROYD, J.G.: Some steady flows of the general elastico-viscous liquid, *Proc. Roy. Soc.* A283, 115–133.

TRUESDELL, C., NOLL, W.: *"The non-Linear Field Theories of Mechanics"*, Springer Verlag.

**[1966]**

CAMERON, A.: *"Principles of Lubrication"*, Longman.

COLEMAN, B.D., MARKOVITZ, H., NOLL, W.: *"Viscometric Flows of non-Newtonian Fluids"*, Springer Verlag.

KAYE, A.: An equation of state for non-Newtonian fluids, *Brit. J. Appl. Phys.* 17, 803–806.

LESLIE, F.M.: Some constitutive equations for anisotropic fluids, *Quart. J. Mech. Appl. Math.* 19, 357–370.

PIPKIN, A.C.: Approximate constitutive equations, in *"Modern Developments in the Mechanics of Continua"*, Ed. S. Eskinazi, Academic Press, 89–108.

**[1967]**

BARLOW, A.J., ERGINSAV, A., LAMB, J.: Viscoelastic relaxation of supercooled liquids, II, *Proc. Roy. Soc.* A298, 481–494.

CLARKE, B.: Rheology of coarse settling suspensions, *Trans. Inst. Chem. Eng.* 45, 251–256.

EDWARDS, S.F.: The statistical mechanics of polymerized material, *Proc. Phys. Soc.* 92, 9–16.

PIPKIN, A.C., OWEN, D.R.: Nearly viscometric flows, *Phys. Fluids* 10, 836–843.

SCHWARZL, F.R., STRUIK, L.C.E.: Analysis of relaxation measurements, *Adv. Mol. Relaxation Proc.* 1, 201–255.

**[1968]**

BERRY, G.C., FOX, T.G.: The viscosity of polymers and their concentrated solutions, *Adv. Polym. Sci.* 5, 261–357.

BROADBENT, J.M., KAYE, A., LODGE, A.S., VALE, D.G.: Possible systematic error in the measurement of normal stress differences in polymer solutions in steady shear flow, *Nature* 271, 35.

CHENG, D.C-H.: The effect of secondary flow on the viscosity measurement using the cone-and-plate viscometer, *Chem. Eng. Sci.* 23, 895–899.

COGSWELL, F.N.: Rheology of polymer melts under tension, *Plastics & Polym.* 36, 109–111.

DAVIS, S.S., DEER, J.J., WARBURTON, B.: A concentric cylinder air turbine viscometer, *J. Sci. Instrum. (Ser. 2)* 1, 933–936.

FARRIS, R.J.: Prediction of the viscosity of multimodal suspensions from unimodal viscosity data, *Trans. Soc. Rheol.* 12, 281–301.

MARKOVITZ, H.: The emergence of rheology, *Physics Today* 21, 23–30.

PLAZEK, D.J.: Magnetic bearing torsional creep apparatus, *J. Polym. Sci. A2* 6, 621–638.


[1969]

GLICKSMAN, M. *"Gum Technology in the Food Industry"*, Academic Press.

TANNER, R.I.: Network rupture and the flow of concentrated polymer solutions, *AIChE J.* 15, 177–183.

TANNER, R.I., PIPKIN, A.C.: Intrinsic errors in pressure-hole measurements, *Trans. Soc. Rheol.*, 13, 471–484.


[1970]

ASTARITA, G., NICODEMO, L.: Extensional flow behaviour of polymer solutions, *Chem. Eng. J.* 1, 57–65.

BATCHELOR, G.K.: The stress system in a suspension of force-free particles, *J. Fluid Mech.* 41, 545–570.

DYSON, A.: Frictional traction and lubricant rheology in elastohydrodynamic lubrication, *Phil. Trans. Roy. Soc.* A266, 1–33.

METZNER, A.P., METZNER, A.B.: Stress levels in rapid extensional flows of polymeric fluids, *Rheol. Acta* 9, 174–181.

SHERMAN, P.: *"Industrial Rheology"*, Academic Press.

TANNER, R.I.: Some methods of estimating the normal stress functions in viscometric flows, *Trans. Soc. Rheol.* 14, 483–507.


[1971]

BATCHELOR, G.K.: The stress generated in a non-dilute suspension of elongated particles in pure straining motion, *J. Fluid Mech.* 46, 813–829.

de GENNES, P.G.: Reptation of a polymer chain in the presence of fixed obstacles, *J. Chem. Phys.* 55, 572–579.

GRISKEY, R.G., GREEN, R.G.: Flow of dilatant (shear-thickening) fluids, *AIChE J.* 17, 725–731.


[1972]

BROADBENT, J.M., LODGE, A.S.: Determination of normal-stress differences in steady shear flow, Part 3, *Rheol. Acta* 10, 557–573.

CARREAU, P.J.: Rheological equations from molecular network theories, *Trans. Soc. Rheol.* 16, 99–127.

COGSWELL, F.N.: Converging flow of polymer melts in extrusion dies, *Polym. Eng. Sci.* 12, 64–73.

COGSWELL, F.N.: Measuring the extensional rheology of polymer melts, *Trans. Soc. Rheol.* 16, 383–403.

HOFFMAN, R.L.: Discontinuous behaviour in concentrated suspensions, *Trans. Soc. Rheol.* 16, 155–173.

HUTTON, J.F.: Effect of changes of surface tension and contact angle on normal force measurement with the Weissenberg rheogoniometer, *Rheol. Acta* 11, 70–72.

KRIEGER, I.M.: Rheology of monodisperse latices, *Adv. Colloid Interface Sci.* 3, 111–136.

LODGE, A.S., STARK, J.H.: On the description of rheological properties of viscoelastic continua, Part 2. Proof that Oldroyd's 1950 formalism includes all simple fluids, *Rheol. Acta* 11, 119–126.

MEISSNER, J.: Development of a universal extensional rheometer for the uniaxial extension of polymer melts, *Trans. Soc. Rheol.* 16, 405–420.


[1973]
CHENG, D.C-H.: Some measurements on a negative thixotropic fluid, *Nature* 245, 93–95.
KESTIN, J. SOKOLOV, M., WAKEHAM, W.: Theory of capillary viscometers, *Appl. Sci. Res.* 27, 241–264.


[1974]
ASTARITA, G., MARRUCCI, G.: *"Principles of non-Newtonian Fluid Mechanics"*, McGraw–Hill.
LODGE, A.S.: *"Body-Tensor Fields in Continuum Mechanics"*, Academic Press.
OLIVER, D.R., BRAGG, R.: The triple jet: A new method for measurement of extensional viscosity, *Rheol. Acta* 13, 830–835.
TADMOR, Z., BIRD, R.B.: Rheological analysis of stabilizing forces in wire-coating dies, *Polym. Eng. Sci.* 14, 124–136.


[1975]
DAVIES, J.M., HUTTON, J.F., WALTERS, K.: Measurement of the elastic properties of polymer-containing oils by the jet-thrust technique, in *Polymères et Lubrification*, Colloques Int. du C.N.R.S., 61–67.
HUILGOL, R.R.: *"Continuum Mechanics of Viscoelastic Liquids"*, John Wiley & Sons.
MÜNSTEDT, H.: Viscoelasticity of polystyrene melts in tensile creep experiments, *Rheol. Acta* 14, 1077–1088.
NOLTINGK, B.E.: (Ed) *"The Instrument Manual"*, 5th edition, United Trade Press.
TRELOAR, L.R.G.: *"The Physics of Rubber Elasticity"*, 3rd edition, Clarendon Press, Oxford.
WAKEMAN, R.: Packing densities of particles with log-normal size distribution, *Powder Tech.* 11, 297–299.
WALTERS, K.: *"Rheometry"*, Chapman and Hall.


[1976]
ACIERNO, D., LA MANTIA, F.P., MARRUCCI, G., TITOMANLIO, G.: A non-linear viscoelastic model with structure-dependent relaxation times, *J. non-Newtonian Fluid Mech.*, 1, 125–146.
BINDING, D.M., WALTERS, K.: Elastico-viscous squeeze films, Part 3. The torsional-balance rheometer, *J. non-Newtonian Fluid Mech.* 1, 277–286.
HAN, C.D.: *"Rheology in Polymer Processing"*, Academic Press.
HARRISON, G.: *"The Dynamic Properties of Supercooled Liquids"*, Academic Press.
HUDSON, N.E., FERGUSON, J.: Correlation and molecular interpretation of data obtained in elongational flow, *Trans. Soc. Rheol.* 20, 265–286.
PETRIE, C.J.S., DENN, M.M.: Instabilities in polymer processing, *AIChE J.* 22, 209–236.
TOWNSEND, P., WALTERS, K., WATERHOUSE, W.M.: Secondary flows in pipes of square cross section and the measurement of the second-normal stress difference, *J. non-Newtonian Fluid Mech.* 1, 107–123.


[1977]
BATCHELOR, G.K.: The effect of Brownian motion on the bulk stress in a suspension of spherical particles, *J. Fluid Mech.* 83, 97–117.
BOGER, D.V.: A highly elastic constant-viscosity fluid, *J. non-Newtonian Fluid Mech.* 3, 87–91.
BOGER, D.V.: Demonstration of upper and lower Newtonian fluid behaviour in a pseudoplastic fluid, *Nature* 265, 126–128.
DAVIES, J.M., HUTTON, J.F., WALTERS, K.: A critical reappraisal of the jet-thrust technique for normal stresses, with particular reference to axial-velocity and stress rearrangement at the exit plane, *J. non-Newtonian Fluid Mech.* 3, 141–160.

GRAY, R.W., HARRISON, G., LAMB, J.: Dynamic viscoelastic behaviour of low-molecular-mass polystyrene melts, *Proc. Roy. Soc.* A356, 77–102.

JOHNSON, M.W.Jr., SEGALMAN, D.: A model for viscoelastic fluid behaviour which allows non-affine deformation, *J. non-Newtonian Fluid Mech.* 2, 255–270.

MIDDLEMAN, S.: *"Fundamentals of Polymer Processing"*, McGraw–Hill.

NIELSEN, L.E.: *"Polymer Rheology"*, Marcel Dekker.

PHAN-THIEN, N., TANNER, R.I.: A new constitutive equation derived from network theory, *J. non-Newtonian Fluid Mech.* 2, 353–365.

TURIAN, R., YUAN, T-F.: Flow of slurries in pipelines, *AIChE J.* 23, 232–243.

WALTERS, K., WATERHOUSE, W.M.: A note on kinematical tensors, *J. non-Newtonian Fluid Mech.* 3, 293–296.


[1978]

JENKINS, J.T.: Poiseuille and Couette flows of nematic liquid crystals, *Ann. Rev. Fluid Mech.* 10, 197–219.

LAUN, H.M., MÜNSTEDT, H.: Elongational behaviour of a low density polyethylene melt, Part 1, *Rheol. Acta* 17, 415–425.

QUEMADA, D.: Rheology of concentrated disperse systems, *Rheol. Acta* 17, 643–653.

SCHOWALTER, W.R.: *"Mechanics of non-Newtonian Fluids"*, Pergamon Press.

WHITCOMB, P.J., MACOSKO, C.W.: Rheology of Xanthan gum, *J. Rheol.* 22, 493–505.

WILLEY, S.J., MACOSKO, C.W.: Steady shear rheological behaviour of PVC plastisols, *J. Rheol.* 5, 525–545.


[1979]

de GENNES, P.G.: *"Scaling Concepts in Polymer Physics"*, Cornell University Press.

FOSTER, J., NIGHTINGALE, J.D.: *"A Short Course on General Relativity"*, Longman

PETRIE, C.J.S.: *"Elongational Flows"*, Pitman.

GRAVSHOLT, S.: Physicochemical properties of viscoelastic detergent solutions, in *"Polymer Colloids II"*, ed. R. Fitch, Plenum Press, 405–417.

LESLIE, F.M.: Theory of flow phenomena in liquid crystals, *Adv. Liq. Cryst.* 4, 1–81.

MEWIS, J.: Thixtropy—A general review, *J. non-Newtonian Fluid Mech.* 6, 1–20.

MÜNSTEDT, H.: New universal extensional rheometer for polymer melts. Measurement on a PS sample, *J. Rheol.* 23, 421–436.

WALTERS, K.: Developments in non-Newtonian fluid mechanics—A personal view, *J. non-Newtonian Fluid Mech.* 5, 113–124.

WILLIAMS, R.W.: Determination of viscometric data from the Brookfield RVT viscometer, *Rheol. Acta* 18, 345–359.

WILLIAMS, G.E., BERGEN, J.T., POEHLEIN, G.W.: Rheological behaviour of high resin level plastisols, *J. Rheol.* 23, 591–616.

WINTER, H.H., MACOSKO, C.W., BENNETT, K.E.: Orthogonal stagnation flow, a framework for steady extensional flow experiments, *Rheol. Acta* 18, 323–334.


[1980]

BALL, R., RICHMOND, P.: Dynamics of colloidal dispersions, *J. Phys. Chem. Liquids* 9, 99–116.

BEAZLEY, K.M.: Industrial aqueous suspensions, in *"Rheometry: Industrial Applications"* ed. K. Walters, John Wiley & Sons, 339–413.

BOGER, D.V., DENN, M.M.: Capillary and slit methods of normal stress measurements, *J. non-Newtonian Fluid Mech.* 6, 163–185.

de KEE, D., TURCOTTE, G., CODE, R.K.: Rheology characterisation of time-dependent foodstuffs, in *"Rheology Vol.3: Applications"*, Eds. G. Astarita, et al., Plenum Press, 609–614.

FERGUSON, J., EL-TAWASHI, M.K.H.: The measurement of the elongational viscosity of polymer solutions, in *"Rheology Vol.2: Fluids"*, Eds. G. Astarita, et al., Plenum Press, 257–262.

FERRY, J.D.: *"Viscoelastic Properties of Polymers"*, 3rd edition, John Wiley & Sons.

HUTTON, J.F.: Lubricants, in *"Rheometry: Industrial Applications"*, ed. K. Walters, John Wiley & Sons, 119–177.

IPPOLITO, M.: Rheology of disperse systems—influence of NaCl on the viscous properties of aqueous bentonite suspensions, in *"Rheology Vol.2: Fluids"*, Eds. G. Astarita et al., Plenum Press, 651–658.

KEENTOK, M., GEORGESCU, A.G., SHERWOOD, A.A., TANNER, R.I.: The measurement of the second normal stress difference for some polymer solutions, *J. non-Newtonian Fluid Mech.* <u>6</u>, 303–324.

MILLS, P., RUBI, J.M., QUEMADA, D.: Suspension flow, in *"Rheology Vol.2: Fluids"*, Eds. G. Astarita et al., Plenum Press, 639–643.

ONOGI, S., ASADA, T.: Rheology and rheo-optics of polymer liquid crystals, in *"Rheology Vol.1. Principles"*, Eds. G. Astarita et al., Plenum Press, 127–147.

VINOGRADOV, G.V., MALKIN, A.Ya.: *"Rheology of Polymers*, Mir, Moscow; English translation, Springer Verlag.

WALTERS, K.: (ed) *"Rheometry: Industrial Applications"*, John Wiley & Sons.

WALTERS, K., BARNES, H.A.: Anomalous extensional-flow effects in the use of commercial viscometers, in *"Rheology Vol.1: Principles"*, Eds. G. Astarita et al., Plenum Press, 45–62.

WALTERS, K., BROADBENT, J.M.: *"Non-Newtonian Fluids"*, Film/Video (U.C.W., Aberystwyth).

WHITE, J.L.: Molten polymers, in *"Rheometry: Industrial Applications"*, Ed. K. Walters, John Wiley & Sons, 209–280.

WHORLOW, R.W.: *"Rheological Techniques"*, John Wiley & Sons.


[1981]

BARNES, H.A.: *"Dispersion rheology: 1980"*, Royal Soc. of Chem., Industrial Division, London.

BRYDSON, J.A.: *"Flow Properties of Polymer Melts"*, 2nd edition, George Godwin, London.

COGSWELL, F.N.: *"Polymer Melt Rheology"*, George Godwin, London.

GALVIN, G.D., HUTTON, J.F., JONES, B.: Development of a high-pressure, high-shear-rate capillary viscometer, *J. non-Newtonian Fluid Mech.* <u>8</u>, 11–28.

MEISSNER, J., RAIBLE, T., STEPHENSON, S.E.: Rotary clamp in uniaxial and biaxial extensional rheometry of polymer melts, *J. Rheol.* <u>25</u>, 1–28.

MÜNSTEDT, H., LAUN, H.M.: Elongational properties (and the recoverable strain in the steady state of elongation) and molecular structure of polyethylene melts, *Rheol. Acta* <u>20</u>, 211–22.


[1982]

BIRD, R.B.: Polymer kinetic theories and elongational flows, *Chem. Eng. Commun.* <u>16</u>, 175–187.

GIESEKUS, H.: A simple constitutive equation for polymer fluids based on the concept of deformation-dependent tensorial mobility, *J. non-Newtonian Fluid Mech.* <u>11</u>, 69–109.

DEALY, J.M.: *"Rheometers for Molten Plastics"*, Van Nostrand Rheinhold.

JONES, W.M., REES, I.J.: The stringiness of dilute polymer solutions, *J. non-Newtonian Fluid Mech.* <u>11</u>, 257–268.


[1983]

ALINCE, B., LEPOUTRE, P.: Flow behaviour of pigment blends, *TAPPI J.* <u>66</u>, 57–60.

GIESEKUS, H.: Disperse systems: dependence of rheological properties on the type of flow with implications for food rheology, in *"Physical Properties of Foods"*, Ed. R. Jowitt et al., Applied Science Publishers, Chap. 13.

JANESCHITZ-KRIEGL, H.: *"Polymer Melt Rheology and Flow Birefringence"*, Springer Verlag.

MEISSNER, J.: Polymer melt rheology—a challenge for the polymer scientist and engineer, *IUPAC Macro-83 Bucharest, Romania*, Plenary and invited lectures, Part 2, IUPAC Macromolecular Division, 203–226.

PRILUTSKI, G., GUPTA, R.K., SRIDHAR, T., RYAN, M.E.: Model viscoelastic liquids, *J. non-Newtonian Fluid Mech.* <u>12</u>, 233–241.

SAUT, J.C., JOSEPH, D.D.: Fading memory, *Arch. Rat. Mech. Anal.* 81, 53–95.
SCHÜMMER, P., TEBEL, K.H.: A new elongational rheometer for polymer solutions, *J. non-Newtonian Fluid Mech.*, 12, 331–347.
WALTERS, K.: The second normal stress difference project, *IUPAC Macro-83 Bucharest, Romania*, Plenary and invited lectures, Part 2, IUPAC Macromolecular Division, 227–237.

[1984]

CROCHET, M.J., DAVIES, A.R., WALTERS, K.: *"Numerical Simulation of non-Newtonian Flow"*, Elsevier.
GIESEKUS, H.: On configuration dependent generalized Oldroyd derivatives, *J. non-Newtonian Fluid Mech.* 14, 47–65.
GUPTA, R.K., SRIDHAR, T.: A novel elongational viscometer, in *"Advances in Rheology, Vol. 4: Applications"* Eds. B. Mena et al., Univ. Nacional Autonom. de Mexico, 71–76.
JACKSON, K.P., WALTERS, K., WILLIAMS, R.W.: A rheometrical study of Boger fluids, *J. non-Newtonian Fluid Mech.* 14, 173–188.
OLDROYD, J.G.: An approach to non-Newtonian fluid mechanics, *J. non-Newtonian Fluid Mech.* 14, 9–46.
PHAN-THIEN, N.: Squeezing a viscoelastic liquid from a wedge: an exact solution, *J. non-Newtonian Fluid Mech.* 16, 329–345.

[1985]

ALVAREZ, G.A., LODGE, A.S., CANTOW, H.-H.: Measurement of the first and second normal stress differences: correlation of four experiments on a poly-isobutylene/dekalin solution D1, *Rheol. Acta* 24, 368–376.
BARNES, H.A., WALTERS, K.: The yield stress myth? *Rheol. Acta* 24, 323–326.
BIRD, R.B.: From molecular models to constitutive equations, in *"Viscoelasticity and Rheology"*, Eds. A.S. Lodge, M. Renardy, J.A. Nohel, Academic Press, 105–123.
CHAPOY, L.L.: (Ed.) *"Recent Advances in Liquid Crystalline Polymers"*, Elsevier Appl. Sci.
de KRUIF, C.G., van IEVSEL, E.M.F., VRIJ, A., RUSSEL, W.B.: Hard sphere colloidal dispersions: Viscosity as a function of shear rate and volume fraction, *J. Chem. Phys.* 83, 4717–25.
GIESEKUS, H.: A comparison of molecular and network-constitutive theories for polymer fluids, in *"Viscoelasticity and Rheology"* Eds. A.S. Lodge, M. Renardy, J.A. Nohel, Academic Press, 157–180.
KELLER, A., ODELL, J.A.: The extensibility of macromolecules in solution; a new focus for macromolecular science, *Colloid Polym. Sci.* 263, 181–201.
MEISSNER, J.: Experimental aspects in polymer melt elongational rheometry, *Chem. Eng. Commun.* 33, 159–180.
ODELL, J.A., KELLER, A., MILES, M.J.: Assessment of molecular correctedness in semi-dilute polymer solutions by elongational flow, *Polymer* 26, 1219–1226.
PEARSON, J.R.A.: *"Mechanics of Polymer Processing"*, 2nd edition, Elsevier, Barking.
SOSKEY, P.R., WINTER, H.H.: Equiaxial extension of two polymer melts: polystyrene and low density polyethylene, *J. Rheol.* 29, 493–517.
TANNER, R.I.: *"Engineering Rheology"*, Clarendon Press, Oxford.
WALTERS, K.: Overview of macroscopic viscoelastic flow, in *"Viscoelasticity and Rheology"*, eds. A.S. Lodge, M. Renardy, J.A. Nohel, Academic Press, 47–79.
WILLIAMS, P.R., WILLIAMS, R.W.: On the planar extensional viscosity of mobile liquids, *J. non-Newtonian Fluid Mech.* 19, 53–80.
WOODCOCK, L.V.: Molecular dynamics and relaxation phenomena in glasses, *Proceedings of a Workshop on Glass Forming Liquids*, Ed. Z.I.P. Bielefeld, Springer Lecture Series in Physics, Vol. 277, 113–124.

[1986]

DOI, M., EDWARDS, S.F.: *"The Theory of Polymer Dynamics"*, Clarendon Press, Oxford.
HEYES, D.M.: Non-Newtonian behaviour of simple liquids, *J. non-Newtonian Fluid Mech.* 21, 137–155.

JONES, W.M., WILLIAMS, P.R., VIRDI, T.S.: The elongation of radial filaments of a Boger fluid on a rotating drum, *J. non-Newtonian Fluid Mech.* 21, 51–64.

LAUN, H.M.: Prediction of elastic strains of polymer melts in shear and elongation, *J. Rheol.* 30, 459–501.

MARRUCCI, G.: The Doi–Edwards model without independent alignment, *J. non-Newtonian Fluid Mech.* 21, 329–336.

MARRUCCI, G., GRIZZUTI, N.: The Doi–Edwards model in slow flows. Predictions on the Weissenberg effect, *J. non-Newtonian Fluid Mech.* 21, 319–328.

PAL, R., BHATTACHARYA, S.N., RHODES, E.: Flow behaviour of oil-in-water emulsions, *Can. J. Chem. Eng.* 64, 3–10.

RAMAMURTHY, A.V.: Wall slip in viscous fluids and the influence of materials of construction, *J. Rheol.* 30, 337–357.


[1987]

BARNES, H.A., EDWARDS, M.F., WOODCOCK, L.V.: Applications of computer simulations to dense suspension rheology, *Chem. Eng. Sci.* 42, 591–608.

BINDING, D.M., WALTERS, K., DHEUR, J., CROCHET, M.J.: Interfacial effects in the flow of viscous and elasticoviscous liquids, *Phil. Trans. Roy. Soc.* A323, 449–469.

BIRD, R.B., ARMSTRONG, R.C., HASSAGER, O.: *"Dynamics of Polymeric Liquids", Vol. 1, Fluid Mechanics*, 2nd edition, John Wiley & Sons.

BIRD, R.B., CURTISS, C.F., ARMSTRONG, R.C., HASSAGER, O.: *"Dynamics of Polymeric Liquids", Vol. 2, Kinetic Theory*, 2nd edition, John Wiley & Sons.

BOGER, D.V.: Viscoelastic flows through contractions, *Ann. Rev. Fluid Mech.* 19, 157–182.

FULLER, G.G., CATHEY, C.A., HUBBARD, B., ZEBROWSKI, B.E.: Extensional viscosity measurements for low-viscosity fluids, *J. Rheol.* 31, 235–249.

GOODWIN, J.: The rheology of colloidal dispersions, in *"Solid/Liquid Dispersions"* Ed. Th. Tadros, Academic Press, Chap. 10.

HUNTER, R.J.: *"Foundations of Colloid Science"*, Vol. 1, Oxford University Press, Oxford.

JONES, D.M., WALTERS, K., WILLIAMS, P.R.: On the extensional viscosity of mobile polymer solutions, *Rheol. Acta* 26, 20–30.

LEONOV, A.I.: On a class of constitutive equations for viscoelastic liquids, *J. non-Newtonian Fluid Mech.* 25, 1–59.

LODGE, A.S., AL-HADITHI, T.R.S., WALTERS, K.: Measurement of the first normal-stress difference at high shear rates for a polyisobutylene/decalin solution "D2", *Rheol. Acta* 26, 516–521.

MARCHAL, J.M., CROCHET, M.J.: A new mixed finite element for calculating viscoelastic flow, *J. non-Newtonian Fluid Mech.* 26, 77–114.

MATHYS, E.F., SABERSKY, R.H.: Rheology, friction and heat transfer study of a discontinuously shear-thickening antimisting polymer solution, *J. non-Newtonian Fluid Mech.* 25, 177–196.


[1988]

AL-HADITHI, T.S.R., WALTERS, K., BARNES, H.A.: The relationship between the linear (oscillatory) and steady shear properties of a series of polymeric systems, in *Proceedings of the 10th International Congress on Rheology Vol. 1,*, Ed. P.H.T. Uhlherr, Australian Society of Rheology, 137–139.

BINDING, D.M.: An approximate analysis for contraction and converging flows, *J. non-Newtonian Fluid Mech.* 27, 173–189.

DEALY, J.M., GIACOMIN, A.J.: Sliding-plate and sliding-cylinder rheometers, in *"Rheological Measurement"*, Eds. A.A. Collyer, D.W. Clegg, Elsevier.

LODGE, A.S.: Normal stress differences from hole-pressure measurements, in *"Rheological Measurement"*, Eds. A.A. Collyer, D.W. Clegg, Elsevier.

MARRUCCI, G.: Paper presented at *Int. Congress on Extensional Flow*, Chamonix, France, Jan. 1988.

MOAN, M., MAGUEUR, A.: Transient extensional viscosity of dilute flexible polymer solutions, *J. non-Newtonian Fluid Mech.* 30, 343–354.

WALTERS, K.: Editorial; Int. Congress on Extensional Flow, Chamonix, France, Jan. 1988, *J. non-Newtonian Fluid Mech.* 30, 97–98.

WALTERS, K., JONES, D.M.: Extension-dominated flows of polymer solutions with applications to EOR, in *Proceedings of the 10th International Congress on Rheology, Vol. 1*, Ed. P.H.T. Uhlherr, Australian Society of Rheology, 103–109.

[1989]

BARNES, H.A.: Review of shear-thickening of suspensions, *J. Rheol.* 33, 329–366.

LAUN, H.M., SCHUCH, H.: Transient elongational viscosities and drawability of polymer melts, *J. Rheol.* 33, 119–175.

# AUTHOR INDEX

(Numbers in *italics* refer to entries in the
Reference section   i.e. pp *171 - 180* )

182

# SUBJECT INDEX

(Numbers in *italics* refer to entries in the
Glossary     i.e. pp *159 - 169* )